Environment and Conflict

Environmental degradation is a fast-growing problem that not only threatens to erode future development and undermine economic prosperity, but also victimises and displaces ordinary peoples and communities in some of the most fragile areas of the world. Often grassroots opposition and mobilisation is seen through a secular lens, implying that collective action is merely material and provincial.

In this book John Agbonifo argues for a decolonisation of the environment and to see the environment from the perspective of local communities. He examines the case of the Ogoni struggle against the Shell oil company, and asks how may we understand the struggle of the Ogoni against the State and Shell? Was the conflict merely about a minority ethnic drive at securing provincial advantages in distributional matters, or the legitimate actions of a local community aimed at preserving its environment and livelihood? Exploring the material and symbolic, provincial and nationalist dimensions of Ogoni motivation, Agbonifo's book is the first serious attempt to discuss these issues.

The book will appeal to scholars and students of the Niger Delta conflicts, resource-related conflicts and social movements, in Africa and elsewhere. Those researching in the fields of development studies, political geography, civil society and collective action will also find it useful.

John Agbonifo is a senior lecturer at the Osun State University, Osogbo, Nigeria.

Transforming Environmental Politics and Policy

Series editors

Timothy Doyle, Keele University, UK and University of Adelaide, Australia and Philip Catney, Keele University, UK

The theory and practice of environmental politics and policy are rapidly emerging as key areas of intense concern in the first, third and industrializing worlds. People of diverse nationalities, religions and cultures wrestle daily with environment and development issues central to human and non-human survival on the planet Earth. Air, Water, Earth, Fire. These central elements mix together in so many ways, spinning off new constellations of issues, ideas and actions, gathering under a multitude of banners: energy security, food sovereignty, climate change, genetic modification, environmental justice and sustainability, population growth, water quality and access, air pollution, mal-distribution and over-consumption of scarce resources, the rights of the non-human, the welfare of future citizens-the list goes on.

What is much needed in green debates is for theoretical discussions to be rooted in policy outcomes and service delivery. So, while still engaging in the theoretical realm, this series also seeks to provide a 'real world' policy-making dimension. Politics and policy making is interpreted widely here to include the territories, discourses, instruments and domains of political parties, non-governmental organizations, protest movements, corporations, international regimes, and transnational networks.

From the local to the global-and back again-this series explores environmental politics and policy within countries and cultures, researching the ways in which green issues cross North-South and East-West divides. The 'Transforming Environmental Politics and Policy' series exposes the exciting ways in which environmental politics and policy can transform political relationships, in all their forms.

For more information about this series, please visit: https://www.routledge.com/politics/series/ASHSER-1371

Global Environmental Governance, Civil Society and Wildlife
Birdsong After the Storm
Margi Prideaux

Environment and Conflict
Place and the Logic of Collective Action in the Niger Delta
John Agbonifo

Environment and Conflict

Place and the Logic of Collective Action in the Niger Delta

John Agbonifo

Routledge
Taylor & Francis Group

LONDON AND NEW YORK

First published 2019
by Routledge
2 Park Square, Milton Park, Abingdon, Oxon OX14 4RN

and by Routledge
52 Vanderbilt Avenue, New York, NY 10017

First issued in paperback 2020

Routledge is an imprint of the Taylor & Francis Group, an informa business

British Library Cataloguing-in-Publication Data
A catalogue record for this book is available from the British Library

Library of Congress Cataloging-in-Publication Data
A catalog record has been requested for this book

ISBN 13: 978-0-367-58866-3 (pbk)
ISBN 13: 978-1-4094-3733-8 (hbk)

Typeset in Times New Roman
by Taylor & Francis Books

Contents

To Joy, Erhan and Nosazena

1 Environment and conflict

Introduction

What drives conflict between local communities, the State and multinational corporations over the environment? Two competing approaches to the problem begin from first principles, delineating how we should understand the environment. Based on such definition, surrogates explain why conflict emerges over the environment without bothering whether their definition is commonly shared by both communities and capitalists. Largely, the definition of the environment on offer is rooted in a modernist scientific worldview, which dispenses with other worldviews. Yet, the delegitimated worldviews remain valid or coherent to local communities. This book focuses on the incommensurability between 'environment' as seen by capital and communities and its role in the emergence of political conflicts in resource-rich domains.

It is not clear in many accounts what environment means. Conventional understanding of environment, foregrounded in a 'technological logic', designates the latter as a resource for human use and background against which social actions are enacted.[1] Champions of this view often include the State, multinational corporations in the extractive sector, and some scholars and journalists. The other vision comprehends environment, beyond its resource value, as a meaningful place, the physical extension of community, and the abode of ghosts, spirits and deities. The worldview rests on a 'moral order' in which environment is not separate from social actors; it is the meeting place of deities, ancestors, human beings and other living and non-living things.[2]

The two logics are inherently conflictual, assuming overt and deadly dimension, wherever industrial capital meets with land-dependent grassroots communities. Everywhere, the story is the same; from the Bergama Movement (Turkey), Naxalite Movement, and Narmada Bachao Andolan Movement (India), the Zapatistas (Mexico), and the Mapuche (Chile), to various grassroots movements in the Niger Delta. The confrontation between the two logics has generated wicked problems for local livelihoods, untold human misery and expropriation. Yet, many scholars continue to use the concepts of 'environment' and 'conflict' without staking out in clear terms what they mean.

The conventional approach shapes scholarship dedicated to the nexus between environment and conflict. Surrogates argue that competition between groups over scarce environmental resources within the context of depleting resources automatically generates violent conflict. Examples such as the Darfur in Sudan and Lake Chad in Northeast Nigeria have been cited to exemplify how scarcity of environmental resources shapes the onset and protraction of violent conflicts in many domains worldwide. A variant of the resource-scarcity school is the resource abundance approach, which suggests that resource abundance does also precipitate violent conflict. The rebel movement, Revolutionary United Front (RUF), is cited as an example of an armed actor that emerged in the context of diamond-abundant Sierra Leone. Greed compelled the RUF's efforts to seize and loot the diamond resource base.

The political economy literature, which focuses more on explaining the socioeconomic causes of environment-related conflicts, has equally been shaped by the conventional understanding of environments. For instance, some argue that the conflicts in the Niger Delta are simply results of the arm-twisting strategy of conflict entrepreneurs who seek cooptation into the State's distributive or patronage network (Reno 2002, 2005; Collier 2001, 2002; Omeje 2006). Others suggest otherwise, arguing that collective mobilisations against oil development and the State reflect a legitimate demand by ethnic minorities for a wholesome environment and secured livelihoods (Obi 2006).

There is, therefore, a marked polarisation within the literature on how best to understand the conflicts vis-à-vis the environment. Like the resource-scarcity and resource-abundant paradigms, the twin political economy approaches to the Niger Delta conflicts limit their discourse of environment to the latter's material and economic value. Inevitably, they simplify the conflicts as strategic attempts by conflict entrepreneurs to capture the material and economic values of environmental resources. Both approaches disable any capacity to understand the environment as more than physical resource. Draped in the garb of the rational actor model, these explanations leave little room for moral motivation as an explanatory factor in conflict.

A view of the environment as synonymous with the material and economic alone represents a Western imaginary. Whitt and Slack emphasise that an 'environment particularises or contextualises a community, situating it within and bonding it to both the natural world and the larger "containing" society' As a result, communities are bounded by and interpenetrated by given environments. Communities transform and partially construct the environment through social practices and discourses. Simultaneously, communities are in turn transformed and partially constructed by the environment. In that sense, environment is 'the embodiment, or material extension of communities'. Communities and their 'constitutive environments are inseparable; they are the unit of development and change. All development is, for better or worse, co-development of communities and environments' (1994: 21–2).

The implication of the argument, which has largely been neglected in the social sciences, is the question of how environmental transformations have shaped societal transformation. The question is placed in proper perspective in the environmental humanities where the quest is to understand how human society shapes the environment, and how the latter, in turn, shapes the human environment, and with what consequences. If environment and society are understood as inseparable, how does environmental transformation imposed by translocal capital or external actors affect the nature and behaviour of the local community? The human dimension of environmental change has focused mainly on the poverty and health impacts. Less attention has been given to how environmental change is implicated in the sudden transformation of a quiescent community into a politically conscious and disruptive one. This book is interested in exploring the environmental history of conflict and collective action in Ogoniland.

What is conflict?

Essentialist conceptualisation of conflict over the environment is a symptom of a deeper malaise, the tendency to reduce the State merely to a political, economic and physical organisation. As a result, instances of the State's projection of symbolic power are obscured. Symbolic power largely facilitates the exercise of political, economic and physical power. The State is a cultural or symbolic organisation (Taylor 1994; Loveman 2005). Failure to grasp the symbolic dimension of the State invariably leads to failure to perceive the symbolic in the challenge that is posed to the State. In consequence, many writers engage in reductionism when they limit oppositional challenge against the State to the level of the economic, political and physical. They define both the State and its challenge aculturally. Contrary to the trend in the larger part of the literature on the Niger Delta conflicts, the author is of the view that a stipulation of what is meant by 'conflict' is appropriate. There is the risk of essentialising reality in the failure to define conflict. Often, there is a tendency to view conflict in the region either as the outcome of durable structural contradictions or structural crisis, or provincial or self-oriented desire of poor marginalised minorities. From the former view, the conflict is seen as an effect of the environmental and socioeconomic crisis, while in the latter the conflict is an expression of shared interests within a common structural condition. In effect, conflict is portrayed as a feature of the social system, or a result of individual beliefs or reaction to external conditions. Structural, or unwelcome conditions, however, do not invariably breed conflict.

For our purpose, therefore, a conflict exists where there is a clear delineation of the identity of the actor, and a definition of the opponent, who are organised and oriented according to their own set of goals and values, and the stakes over which they fight (Touraine 1985). It is a relationship between two or more opposed actors 'fighting for the same resources, to which both give value' (Melucci 1985: 794). Actors in conflict reveal the stakes in conflict. They do not simply fight for material provincial goals. Their goal is not merely to be included in the status quo. They desire to change people's lives in the belief that the individual's life can be changed in the present while pursuing more general changes in the society. Thus, actors in conflict fight for symbolic or cultural stakes; 'for a different meaning and orientation of social action' (Melucci 1985: 797).

There are different types of social conflict, including collective efforts to advance group interest in a competitive situation, change the rules of the game for group benefit, and control of the main cultural patterns, that is who defines the truth, what constitutes morality and how people relate with the environment. Moreover, society is composed of different sub-systems to which the collective actor is oriented. A collective actor functions simultaneously within several organisational systems. In consequence, her/his activities connect with a whole range of objectives, problems and actors (Touraine 1981). Failure to grasp the complexity of the collective actor and the arena in which it operates has led some to deny, unwittingly, dimensions of a conflict that involve control of the 'main cultural pattern', and to reduce all conflicts to the presence of grievances, mundane aspirations, and political claims.

When does conflict emerge in the form of collective mobilisation? Is it during a time of crisis or structural dislocation? A view of collective action in the Niger Delta as reaction against structural disintegration or crisis hides the conflict dimension of the collective actor. Actors are deprived the meanings of their action (see Melucci 1985). Similarly, the suggestion that conflict is the expression of shared interests or beliefs does not clarify how interests and beliefs are produced in the context of constraints. By limiting collective action to the political level, we hide their cultural dimension. Social conflicts are not simply political; they are cultural as well given that they shape the system's cultural production. Collective action is not simply about inclusion, it 'challenges the logic governing production and appropriation of social resources' (Melucci 1985: 798).

Largely, the literature on the conflict has underlined the causative role of environmental crisis, economic crisis, and political instability. What has been neglected is the role of culture in the emergence of the conflict. Culture has been treated as insignificant background to the conflict. Thus, the conflict is seen as emergent and disconnected from its institutional contexts. The argument is taken to extreme by Bob (2005) who strenuously argue that Ogoni frames were shaped by global factors. This book argues that there is continuity between the Ogoni mobilisation and its institutional contexts. The challenge Ogoni mobilisation generated and the frames it deployed emerged from within its cultural universe, even if it borrowed from the global space.

Constructing the environment

In the cosmology of Niger Delta communities, the environment is more than resources and far from an inanimate stage for human practices. To the contrary, the environment represents the encounter between the human, other-worldly, and other animate and inanimate beings, including the veneration that accompanies that interaction. Desecration of that linkage elicits moral sentiments of revulsion and unease. Chakrabathy (2000:16) underscores the interconnection between humans and environment, arguing that the ontological assumption embedded in the social sciences that 'the human is ontologically singular, that gods and spirits are in the end "social facts," that the social somehow exists prior to them' is misleading. There is no known society in which humans have existed without gods and spirits associating with them. To him, gods and spirits are 'existentially coeval with the human', and 'the question of being human involves the question of being with gods and spirits' (Chakrabathy, 2006:16). Such companionship takes various forms in, on and within the environment.

Communities are, however, just one set of actors who construct, and are constructed by, the environment. Through policies and laws, the State ensures the spatial ordering of a region into reserved areas, crude oil pipelines, and housing settlements. Moreover, the State employs its symbolic power in an attempt to naturalise its contingent processes of spatialisation. Thus, the materialistic tools of laws and policies have cultural dimensions (Loveman 2005). Similarly, oil corporations extend spatial ordering by means of land enclosure, land and river pollution, the siting of oil facilities, and the discursive redefinition of local places in line with extra-local logics.

The State is more than an administrative, policing and military organisation; it is also a symbolic organisation (Loveman 2005). Symbolic power is the 'ability to make appear as natural, inevitable, and thus apolitical, that which is a product of historical struggle and human invention' (Loveman 2005: 1655). Through classification, codification and regulation, the State is able to constitute particular kinds of people, places and things. Through spatial practices and policies the State actively constitutes the people and places in whose name or interest it claims to exist or act. Social life is ordered through generalised perception that a State's activities are natural, inevitable or manifestly useful.

The spatial orderings disrupt the pre-existing and on-going spatial logics and practices of local communities. Therefore, when local communities mobilise against the State and oil corporations, environmental construction or place-making become the object of contentious politics. Such mobilisation reflects utter spatial transgression, disrupting the 'official' orderings of the State and oil corporations in an attempt to reclaim pre-colonial rights to territorialisation. The conflict challenges the State's symbolic power, or the naturalisation of State's constitution of the environment, and the notion that its practices are inevitable and for the general good.

Grassroots approach to environment

The modernist perspective recognises only what can be derived from empirical experience. Knowledge that is not empirically tested, such as indigenous knowledge, is seen as illegitimate and disqualified as inadequate and unscientific

(Fairhead and Leach 1997: 54). By recognising scientifically verified knowledge, modernist epistemology marginalises critical lines of inquiry (Preston 2002). Instead of empirical evidence being the source of all knowledge, the mind is active in the construction of knowledge; human beings do not discover knowledge but they construct it (Schwandt 2002). This book adopts the decolonised approach to environment. The author believes that rather than being objective, knowledge is 'engaged, value bound and context determined' (Scoones and Thompson 1993: 9). Also, arguments that research is an independent process of discovering the truth do not reflect reality (Finlay 2002). The exercise of power shapes and validates some types of knowledge claims over others (Radford 1992). Despite their weaknesses, qualitative method and use of unstructured interviews, follow-up interviews, in-depth group discussions and anecdotal evidences provide access to local community subjective reality that would otherwise remain hidden under a positivist approach.

Place-sensitive social movement approach

The central part of the book examines the potential of place-sensitive social movement theories for understanding the Ogoni conflict, and why and how the Ogoni mobilised against the State and Shell Oil Company. The idea is to analyse how place, that is, environment understood from the perspective of local communities, mediated why and how the community mobilised. Existing literature on the conflict provides robust account of factors implicated in the conflict. Yet, by ignoring place-sensitive social movement imagination, the literature is clearly unable to explain who joined the movement, or did not, and why. It is more than a question of why the movement emerged; but more of why specific people decided to join the movement in particular places and times. Place-sensitive social movement imagination enables engagement with such spatially precise, actor-oriented question by directing attention to a deeper understanding of how the movement came into being (Wolford 2003).

Place-sensitive movement approach enables us overcome the duality between society and nature or the human and non-human inherent in many accounts. The conventional understanding of environment is related to the basic sociological premise that 'social facts' are to be explained by 'social facts' (Lockie 2004). Thus, environment-related movements and the aspirations of activist members have been conceived in terms of social psychological causes and processes. In effect, the environment is reduced to asocial and passive entity, whose significance is limited to economic value. Reconceptualisation of the environment as constructed, and therefore a meaningful place, helps overcome such ontological distinction between the human and non-human. Place-sensitive movement theories provide the tools to navigate the interrelationship and interdependence between the human and non-human.

Place-sensitive social movement theories enable better understanding of the complex realities of conflicts. The conflict is often defined in terms of crude oil, or the precipitating issue. Similarly, the goals and intentions of the protagonists are derived ab initio from the economic importance of crude oil and what value it entails for the conflicting parties. In effect, conflict that emerges over crude oil is invariably about protagonists desire to appropriate the benefit or market value of oil. Environmental protection discourse of the people is merely a metaphor for an attempt to extract greater value from crude oil revenue. Invariably, the issue at stake defines what conflict is – competition between the contenders to improve their lot vis-à-vis the issue. Such understanding of conflict reduces conflict to a single thing usually geared at material appropriation. It disables analytical focus on the multi-dimensionality of conflict, the various societal subsystems conflict is directed at, and the complex forms of challenge posed by conflict. It is unable to account adequately for why and how collective actors mobilise.

The lack of deserved attention to the role of place in the onset and dynamics of conflicts in the region has promoted the proliferation of con-tinental resolution measures even though no two conflicts are exactly the same. The author believes that continental perspectives to understanding the conflicts in the region must be abandoned for a place-sensitive approach. Too many studies of the conflict deploy continental perspectives that explain one conflict and not another. Continental perspectives hide the diversities in causes and courses of conflicts (Shaw 2003). Thus, many works on the conflict are based ab initio on unfounded and irrelevant arguments. Too many of the books on the conflict are aspatial in intent and/or execution. They fail mis-erably to understand the environment from a decolonised perspective, that is, as place. That is why place-sensitive social movement theories provide a turning point for this book.

This book is the outcome of about ten years' experience of researching the Ogoni conflicts. The author has had in-depth conversations with many past and present Movement for the Survival of Ogoni People (MOSOP) leaders, activists, ordinary Ogoni men and women around the issue of why and how they mobilised. The author has published widely on the Ogoni conflict and believes that the insights derived from this research have wider and general applicability. Importantly, the social movement approach of this book broadens social movement scholarship in Africa, and bridges the deficit in movement scholarship, compared with North America, Western Europe and Latin America, in the continent.

What is place?

In many accounts on oil and conflicts in the Niger Delta, the latter is taken as synonymous with the environment, perhaps explaining why little attempt is made to define either what environment or the Niger Delta mean. There is a need to fill the gap. The name 'Niger Delta' represents the picturesque branching off of the Niger River into several tributaries across an expanse of land before emptying into the Atlantic Ocean. It is simply a physical geo-graphic concept that gives little attention to people. It is little wonder then that, in recent times, the Niger Delta is synonymous with natural resources, notably crude oil. In the geographic and resource imaginations, the Niger Delta is no more than a piece of land shaped by water bodies and sub-surface rivers of crude oil. The problem with this representation is that the perspective continues to shape the human approach to the environment and oil development in the region. Oil development is rooted in a modern developmental gaze, a technological order that sees the Niger Delta as nothing more than a difficult terrain bristling with abundant resources. The effect is that oil development proceeds with little regard to the interests, values and worldviews of people who inhabit the region.

Now, the Niger Delta as a term has roots in the English language. Before colonial intrusion, there was little idea of a definable geographical space, the Niger Delta. So, what is universally defined as the Niger Delta was composed of different places, bearing differing meanings to the many ethnic groups that inhabited them. Thus, it is arguable that the place the Ogoni inhabited meant something to the Ogoni that was different from what living in Benin City meant to the Binis, and on and on. The variety of places and the differing meanings they embodied have been collapsed into and erased with the centralising concept of the Niger Delta. By approaching the region on the basis of what is common to it, its resource endowment and physical qualities; we are inhibited from seeing the varieties of meanings the various peoples charge different places with.

Therefore, to appreciate the meaning of environment we can decide to go along with the modernist view that came with colonialism or effect a mental decolonisation by exhuming the pre-colonial worldview of environment. Thus, the question becomes, what does the environment mean to the Ogoni, the Ijaw, the Itsekiri, the Binis, and the Urhobos? Again, we are bound to come up with a variety of meanings. That very fact provides a fundamental argument for seeing the environment from the local perspective rather than a universalising, top-down modernist perspective. Local communities partially constructed the region prior to the arrival of capital and its oppositional modernist logic. In the attempt to understand the clash of logics, it is a moral issue that we explore the meanings the environment conveys to the pre-established local communities. The environment, therefore, must be injected with what locals make of it, rather than seen as something external to them. The concept of place enables us bridge the modernist gap between community and environment.

Agnew (2005: 86) defines place as 'the encounter of people with other people and things in space. It refers to how everyday life is inscribed in space and takes on meaning for people and organisations'. Agnew (1987: 28) disentangles the concept of place thus:

> Interwoven in the concept of place ... are three major elements: *locale*, the settings in which social relations are constituted (these can be informal or institutional); *location*, the geographical area encompassing the

settings for social interaction as defined by social and economic processes operating at a wider scale; and *sense of place*, the local 'structure of feeling'.

(1987: 28)

Location captures the physical geographical area and the ways in which economic and political developments, operating on a wider scale, impact on it. The emphasis is on 'macro-order' affects on a place and 'the ways in which certain places are inscribed, affected and subject to the wider workings of economic and political structures that normally originate from outside the area itself' (Oslender 2004: 961). *Locale* refers to the formal and informal arena in which everyday social interactions and relations take place. *Sense of place* refers to the ways in which 'human experience and imagination appropriates the physical characteristics and qualities of geographical location' (ibid.: 962). It stresses how individuals and communities develop attachment to places through experience, memory and intention. Rather than separate rigid entities, it is best to consider the three components of place as entangled (Oslender 2004: 963). Accordingly, any pretensions to place sensibility, which ignores one or more of the three components, fall below the bar as employed here.

In tandem with Agnew's definition of place, local communities' conception of environment embraces the observable physical terrain and the local arena in which formal and informal transactions take place, the social and economic encounters with other people and forces, and the ghosts, values and meanings they invest place with. Formal and informal interactions take place in the context of ghosts, which are seen as inhabiting the physical environment. Environment-related conflicts are essentialised when explained solely in terms of location, that is, grievances arising from impacts of the interactions between local people and external forces. It stands to reason that disaffection is generated when external actors and forces unleash processes that transform the economic and political context of local people, particularly in disadvantageous directions.

More deeply, however, the interaction becomes insidious because it not only imposes unwelcome social and environmental change, but undermines the world as the locals know it. Local understanding of the environment as a harmonious entity populated by plant and animal life, ghosts and ancestors and meanings and values come under unmitigated assault from a modernist perspective ill-equipped to comprehend such a worldview.

Place-sensitive social movement theories

Many social movement theorists primarily treat place as dead geography. Place is understood more as a background to social events and less as constitutive of contentious politics. Although some movement scholars have underlined the spatiality of conflicts, their accounts of space are underspecified because they have barely been able to show that place mediates conflicts, how it unfolds, and that it is often the stake in conflict (Tarrow 2001).

Scholarship on disruptive protest tactics emphasises the ways in which protest tactics are used by movements to upset institutional and social arrangements. Piven and Cloward (1977) reference place when they suggest that the institutional pattern that aggregates and separates people shapes the kinds of protests that can occur. Yet, they show little concern for how protesters use place to challenge and disrupt such patterns (Dikec 2005). Land occupation, sit-ins or barricades are spatial transgressions; they make visible what should not be seen. Such contentious activities transgress and destabilise power relations inherent in dominant social organisations or spatial structures.

Resource mobilisation (RM) theory is concerned with how movements emerge. It contends that movement mobilisation is a function of the availability and deployment of vital resources such as funds, capable leadership, organisational structure, and networks. The perspective give scant attention to the view that what resources a movement is able to access and deploy is shaped by place, and that what a movement considers as key resource is equally a function of place. RM ignores the place-specific cultural symbols, values, meanings and ghosts that collective actors can capitalise on to spark mobilisation and entrench members' commitment.

The New Social Movement perspective focuses on why movements emerge rather than how movements form. In that respect, it locates movement emergence as a reaction to structural transformations occurring in Western societies. The new movements are contrasted with the 'old' movements, which are alleged to be a reaction to bread and butter issues or labour and economic matters. The 'new' movements, in contrast, are concerned with issues of identity, environment, ethics and cultural demands (Melucci 1985). The new movements arise from the realisation that the system that is challenged controls not only the 'means of production' but the production of symbolic goods, that is, of information and images, of culture itself (Touraine 1985).

The perspective privileges discursive struggles or symbolic challenges to dominant codes, and fails to offer a robust conception of place with regard the emergence of contention. It ignores the understanding that discourses and symbols are place-based. Symbols and discourses, which resonate in one place, may be meaningless in another place. The nature of place shapes what symbols and discourse emerge there. The perspective acknowledges the importance of public space in making power visible, but reduces space to committees and task forces (Melucci 1989: 228–229).

Framing theory has also shown little interest in the role of place in social movements. Although framing theory has enabled a shift from the material to the cultural dimension of mobilisation, there is little attempt to emphasise the place-based framing of collective actors (Wilton and Cranford 2002). Framing scholars show how movements build collective identity using 'interpretive schemata' derived from race, class, gender, and environmental injustices (Gamson 1992; McAdam, McCarthy, and Zald 1996; Snow and Benford

1992). But few address the fact that ideational forces, including gods, beliefs and tradition, sit in place and how they shape collective identity and frames.

There is more to a collective actor than resources, political opportunities, and the broad structural context of its emergence, identity, motivations, strategies, tactics and frames. The place in which a movement emerges significantly shapes the structure and dynamics of the movement. Place-based beliefs, history and worldviews shape social movement dynamics and collective identity. Place-sensitive movement theories are those that give attention to how place mediates collective action.

Previous studies and gaps in the literature

Abubakar Momoh, Osaghae, Osadolor, Akinyanju, Oyovbaire, Onwudiwe, Anikpo, and Ikelegbe, have emphasised in various ways the problem of the Niger delta as an issue of 'the National Question'. The core of the National Question relates to how people are organised, empowered or disempowered, which arose from the amalgamation of the Southern and Northern Protectorates, the subsequent incapacity to transform the complex into national societies and the consequent problem of what to do with the country. The contradictions generate anger, frustration and hostility to the State and multinational oil corporations. Such hostility has resulted in forms of violent protest.

Nigeria is a 'rentier space', defined by 'high stake rentier politics', or a political tradition of desperate tendency to accumulation (Omeje 2006). Complementing the structural emphasis are arguments that locate the emergence of conflict in perceptual differences. In this view, State conception of security stands in contradiction to citizens' notion of security, resulting in violence (Ibeanu 1997).

Despite the huge literature on the Niger Delta and its contribution to understanding the Ogoni conflict, significant blind spots remain in our understanding. While the literature is very robust on materialist and structural explanations, it is thin on the cultural basis of the conflict. Invariably, the conflict is enframed as an effect of structural factors. At other times, the conflict is blamed on individual greed or emotions. Culture is taken as background to the conflict, but never as the instigator.

An aspect of culture hardly evident in the literature as a result of the neglect of cultural causality generally, is the role of symbolic power in the conflict. The State and Shell are apprehended as deploying physical force, political and economic power, and subterfuge to control the Ogoni and their environment. Of course, the State's action is merely geared at sustaining its economic and political interest in capital accumulation. It fails to register that the State also reserves symbolic power, and that successful deployment of physical, economic and military power rests to some extent on its exercise of symbolic power. How the State and Shell use symbolic power in that effort is unaddressed.

Inevitably, the challenge to the State is construed in terms of economic and political power alone. The existing literature invariably understands insurgent politics as relating only to gaining material advantages at the expense of the State (Ibeanu 1997; Ikelegbe 2001; Welch 1995; Osaghae 1995b: 333; Omeje 2005). The literature explains Ogoni demands as 'collective goods' meant for the benefit of their constituencies. What is hidden is the cultural and symbolic challenge that the Ogoni challenge poses to the cultural and symbolic power of the State. The theoretical approach presented here charts a different direction by encouraging recognition of the nonmaterial dimensions of Ogoni demands.

The dynamics of the Ogoni movement and the explanation by necessity invites an understanding of the experiences of the latter within the framework of place-informed social movement theory. Existing explanations of the Ogoni movement failed to recognise how place mediated movement agency. However, attention to 'place' gives insight into why the movement emerged where it did, the nature of the movement, and the spirit, or what inspires and motivates individuals within the movement. This book contributes to the literature by adopting a place-sensitive perspective in combination with social movement theory.

In the existing literature, the Ogoni struggle has stagnated in claims and counter-claims about what factors explain the Ogoni rebellion. A place-sensitive social movement approach to the Ogoni movement provides space to open up debate on the struggle. Moreover, the theoretical approach is relevant and timely in the context of Sub-Saharan Africa where there is paucity of social movement scholarship compared to vast literature on the subject in other parts of the world. In effect, a social movement approach to the Ogoni conflict contributes to bridging the theoretical gap as well as building up scholarship on social movements in Africa.

Place and space are mutually constitutive of social movement agency. It is important to know the place where a movement emerges, where the movement activists live and the meaning living in that particular place conveys to them (Oslender 2004). The core of such sensibility is that the 'place and the subjectivities, identities and passions that it generates with locals make a difference to the ways in which a movement organises and articulates itself' (ibid.: 958). A place is constituted by the ghosts we take to inhabit and possess it, ghosts of the dead and living, individuals and collectives, of others and ourselves (Bell 1997: 813).

The social movement approach enables critical engagement with why and how the Ogoni mobilised against the State and Shell in contradistinction to a significant amount of scholarship, which remains content with understanding merely why the Ogoni mobilised. The result is that most scholars adopt a structuralist lens that portrays conflicts 'as the expression of 'underlying' forces, events that could be deduced or even predicted from structural causes' (Starn 1992). This reduces the conflict to a reaction of the Ogoni to these structures. By reducing the Ogoni to mere pawns over-determined by social structure, the structuralist perspective is unable to account for the quiescence of other Niger Delta groups in the face of domination. Moreover, it hides alternative model of development, social organisation and conflict resolution.

Furthermore, writings on the Ogoni movement are inadequate because they suggest a linear link between grievances and collective action, a view partly shaped by the overriding focus on 'why' the Ogoni mobilised. They ignore the cognitive, spatial and micro-mobilisation processes involved in the movement from grievances to mobilisation. Such scholarship is not helpful to a holistic understanding of conflict, conflict prevention and resolution. The current approach details those neglected processes. By so doing, the study provides critical lessons to students and would-be collective actors, and identifies several early-warning signals that may be useful to authorities, development actors and peace activists.

The existing literature invariably understands insurgent politics as relating only to gaining material advantages at the expense of the State (Ibeanu 1997; Ikelegbe 2001; Welch 1995; Osaghae 1995b: 333; Omeje 2005). Thus, the literature unambiguously casts the demands of the political entrepreneurs as 'collective goods' meant for the benefit of their constituencies. The theoretical approach presented here charts a different direction by encouraging recognition that the Ogoni desired restructuring of the polity as a strategy to assure equality to all federating ethnic units. Sensitivity to the nationalist and moral dimensions of the movement has been conspicuously absent in many accounts.

The aim, scope and organisation of this book

Aim and approach

The dynamics of the Ogoni movement and the explanation by necessity invites an understanding of the experiences of the latter within the framework of place-informed social movement theory. Existing explanations of the Ogoni movement failed to recognise how place mediated movement agency. However, attention to 'place' (as a metaphor for the contestation of environmental resources and values) gives insight into why the movement emerges where it does, the nature of the movement, and the spirit, or what inspires and motivates individuals within the movement. The book bridges the gap in the literature by adopting a place-sensitive perspective in combination with social movement theory.

Social movement agency is inseparable from its environmental context. The identities and passions that a place generates with local inhabitants shape how a movement of the locals organises or articulates itself. Every place has its own unique ghosts (Bell 1997).

The social movement approach enables critical engagement with why and how the Ogoni mobilised against the State and Shell in contradistinction to a significant amount of scholarship, which remains content with understanding merely why the Ogoni mobilised. The result is that most scholars adopt a structuralist lens that portrays conflicts as reflection of certain fundamental causes, which flow from and are evident in the social structure (Starn 1992). This reduces the conflict to a reaction of the Ogoni to these structures. By reducing the Ogoni to mere pawns over-determined by social structure, the structuralist perspective is unable to account for the quiescence of other Niger Delta groups in the face of domination. Moreover, it hides alternative modes of development, social organisation and conflict resolution.

Furthermore, writings on the Ogoni movement are inadequate because they suggest a linear link between grievances and collective action, a view partly shaped by the overriding focus on 'why' the Ogoni mobilised. They ignore the cognitive, spatial and micro-mobilisation processes involved in the movement from grievances to mobilisation. Such scholarship is not helpful to a holistic understanding of conflict, conflict prevention and resolution. The current approach details those neglected processes. By so doing, the study provides critical lessons to students and would-be collective actors, and identifies several early-warning signals that may be useful to authorities, development actors and peace activists.

The existing literature invariably understands insurgent politics as relating only to gaining material advantages at the expense of the State (Ibeanu 1997; Ikelegbe 2001; Welch 1995; Osaghae 1995b: 333; Omeje 2005). Thus, the literature unambiguously casts the demands of the political entrepreneurs as 'collective goods' meant for the benefit of their constituencies. The theoretical approach presented here charts a different direction by encouraging recognition that the Ogoni desired restructuring of the polity as a strategy to assure equality to all federating ethnic units. Sensitivity to the nationalist and moral dimensions of the movement has been conspicuously absent in many accounts.

Organisation of the book

The book is organised as follows: Chapter 1 argues that there is little if any gain from the common tendency in the academic debate, which explains the Delta conflicts in either materialist or provincial terms. It examines major theoretical perspectives, which portray the Niger Delta conflict as the legitimate action of local communities, and as no more than a struggle for inclusion in a patronage network propelled by the desire for personal accumulation. It argues that it should be a matter for empirical investigation whether provincial economic, political or nationalist interests or, a combination of forms creates the underbelly of the conflicts. Chapters 2 and 3 provide a brief overview of the context of the Ogoni by exploring the Ogoni context in terms of location and locale respectively. The idea is to explore how extra-local or global economic and political factors structured Ogoni, as well as how the national, regional and cultural contexts shaped the Ogoni environment and mobilisation. Chapter 4 deploys metaphors to explore the contact between capital and the Ogoni. It argues that both actors represented different logics, which inevitably led to the violence that emerged around crude oil exploitation. Chapter 5 and 6 examine why and how the Ogoni mobilised, arguing that the elements of place, location, locale and sense of place, shaped Ogoni mobilisation. Chapter 7 explores the cultural basis of Ogoni mobilisation to open up argument against those who see the conflict as determined by social structure or individual psychology. Chapter 8 focuses on the moral basis of Ogoni mobilisation in contrast to a large portion of the literature, which fingers selfish provincial interests, or materialist considerations as factors that inspired the conflict. Chapter 9 argues that even the success or otherwise of collective mobilisation is shaped by the place where the mobilisation emerged. Chapter 10 is the conclusion.

Notes

1   Technocratic order.
2   Moral order.

2 Context matters

Ogoni and place making

Introduction: historical flows and place-making

This chapter focuses on the geographical, cultural and historical setting of this study. This is important because the space, culture and history of the Niger Delta frame the general subject and specific problems of the book. How we conceptualise the relationship between communities and nature determines the way we see and relate with the world. Traditional communities view their most valuable possessions as land and culture, which are seen as inextricably linked. One cannot exist without the other. Thus, the separation between culture and nature is a modernist thing. There are three definitions of nature; 'nature as a category of "things"; nature as space that is not human; and nature as inner essence' (Pilgrim and Pretty 2010: 4). Modern industrial capital perceives nature and culture as two distinct entities. And in some cases, they are seen as adversaries whose interactions result in a win–lose scenario. Yet, some cultures have a strong sense of attachment to nature.

Many books on the Niger Delta, and the Ogoni, fail to delineate the dynamic relationship between nature and culture. As a result, how human culture shapes biodiversity and the transformation of the Niger Delta landscape is left hidden. One effect of that is the tendency to see the region as pristine. When we understand that the region is the outcome of human and geographical processes, then we can appreciate the region's landscape as a manifestation of the various cultures that compose it. Also, the landscape partly reflects the imprints of global capital and the conflict between its territorialising logic and local community place-making practices. It is, in effect, a 'cultural archive, a record of human endeavour and husbandry' (Adams 1996: 106).

Recent engagement with events in the Niger Delta promotes two erroneous views of the region; prior to the activities of oil corporations, the Niger Delta was a pristine landscape, and the view that the landscape and human communities were separate. The vision of a pristine landscape is unable to appreciate how over time communities actively constructed the region practically and discursively. The separation between nature and society is evident in the grammar of 'pristine nature'; the separation ensures the untouchability of nature. Moreover, the dualism promotes an erasure of the mutual constitution and construction of nature and community. With an untouched nature and a removed human community, it is a short step to the modernist dream of disenchanted natural resources, for which value derives only from pillaging. The overall effect is that most engagements with the region lack an understanding of the historical and geographical production of the region's topography.

This chapter sets the stage for an understanding that the places that compose the Niger Delta are products of ongoing sociohistorical processes. The human dimension of the process of landscape transformation was largely indigenous in pre-colonial times. However, the logic and trajectory of the dynamics of change encountered an extremely opposing logic and culture with the onset of colonialism, and, more important for us here, oil capitalism. In effect, in different epochs and for differing reasons, the region has been a translocal strategic action field where distant and local actors collaborated or competed over whose meaning of the environment counted and how the environment should be used. Both local communities and global capitalists have struggled to construct and use the environment in ways that each deemed beneficial. For this reason, this chapter attempts to explore the dynamic relationship between cultures and nature in time. We begin briefly with the social construction of the larger Niger Delta region.

Location

In history, the Niger Delta has different meanings for different groups of people. The marine topography provided sanctuary for numerous migrant groups escaping the reach of repressive local potentates and a rich source of livelihood for its inhabitants. From the fifteenth century, the Niger Delta was the heart of European contact and trade (in human beings and palm oil) with West Africa. When mutual respect and beneficial trade relations gave way to European colonisation, the region, hitherto a safe haven and centre of growing economic activities, became the theatre of military invasions, punitive expeditions and conquests. Evocative of the resistance against European colonisation and dispossession, the Niger Delta, since the early 1990s, has again become the flashpoint of communal mobilisations and violent state repression. At the centre of the recent conflicts are multinational oil firms whose operations and impacts on the environment, including the role of the State, have come under widespread condemnation (Catholic Secretariat of Nigeria 2006; Catholic Relief Services 2003).

The Niger River is the longest and largest river in West Africa, and the third longest in Africa (the other two are the Nile and the Congo). It is difficult to imagine Nigeria without the Niger (Ifemesia 1982: 25). The total drainage area of the Niger, including its tributaries and outlets spread over some 222,000 square miles, approximately 60 per cent of the total area of the country. Since pre-colonial times, the Niger River has been a major highway of trade and contact between the different people in the region. Traders employed large dugout canoes for long-distance trade along the river. Delta traders went as far upstream as the Niger-Benue Confluence, and Igbira and Hausa traders from the north of the country came as far downstream as Aboh and Asaba (ibid.).

The Niger Delta comprises the area covered by the natural delta of the Niger River and the oil-producing areas to the east and west. Its northernmost boundary is the town of Aboh on the Niger River. The Benin River forms its western boundary and the Imo River its eastern limits (UNDP 2006: 44). To the south, the region's swamp forests are marked off from the Atlantic Ocean by a fringe of mangroves about 10 kilometres inland. The area covers roughly 25,900 square kilometres. For political and administrative reasons, the Niger Delta today consists of nine states (Abia, Akwa Ibom, Bayelsa, Cross River, Delta, Edo, Imo, Ondo and Rivers) and 185 local government areas.

The Niger Delta, some 5,600 square miles or 14,400 square kilometres, is a vast forested wetland. Although it has provided refuge and an environment that supported thriving human and nonhuman species, the Niger Delta has become a theatre of depredation and a subject of controversy. This development relates to the fact that the same sediment that formed the region and supported life on it contains crude oil, an international commodity. According to Berns and Roberts (2002),

> the Niger Delta occupies a vast area crisscrossed by rivers, tributaries, swamps and lagoons. Water, therefore, has always been far more than a mere element of nature. Water is synonymous with life itself, with spiritual sustenance, with wealth and prosperity, and especially with communication and identity. For years, the waterways of the Niger Delta have connected and divided people, serving as conduits and obstacles, repositories of riches and realms of danger. The ambivalence associated with these contrasting potentialities is made manifest through the arts, cultures, and ethos of the many peoples inhabiting this aqueous region.
>
> (Berns and Roberts 2002: 11)

People with diverse cultural traditions coexisted in the Delta, sharing a common, mostly aqueous environment, inundated by floods, tides and tropical rainstorms. Widely distributed resource bases encouraged the use of rivers as medium of communication and commerce. Although most cultural entities maintained their languages, intercourse engenders similar customs. Thus, rather than separate peoples, the numerous waterways of the Delta created cultural convergence (Anderson and Peek 2002). Ethnically diverse, the region's inhabitants represent a number of different language groups, each group composed of distinct languages, many of which are not mutually intelligible. The five main 'linguistic and cultural groups – Ijoid, Edoid, Delta Cross, Yoruboid and Igboid – are each composed of numerous sub-groups' (UNDP 2006: 48). The Ijoid group predominates and many consider the group the longest settled in the region. Many of the languages also contain numerous dialects. Many people in the region speak two or more Nigerian languages in addition to Pidgin English (ibid.: 29). The Ijo speakers find the Ijebu-Yoruba on their far west, on the western and northwestern flank the Itsekiri, Urhobo and Isoko, on the north and northeast the Igbo, and on the east the Ogoni and Ibibio. Yet, the Ijaw language differs from those of neighbours even though it shares with them certain cultural and social features. The Urhobo and Isoko speak dialects of Edo but are culturally distinct from the Itsekiri. Although culturally closer to Benin, the Itsekiri speak a Yoruba language (Leis 2002).

It is against that backdrop that Anderson and Peek (2002: 30) reasoned that perhaps the best conception of the Niger Delta is as a 'conceptual framework. Its inhabitants exist within a unique fabric of cultural resemblances and cultural differences' (Jung 2003: 457).

Beyond facilitating trade and communication, the Niger River performs many other economic and cultural functions. It is a food source for the communities on its bank, and the waters of the Niger help irrigate farmlands. The Ijo use the river for many activities including fishing. Fish is an important part of the traditional Nigerian diet. The Niger is also a space for recreation and sport for the Delta people as evidenced by the annual Pategi regatta and the Argungu festival. In addition, Ifemesia shows that the river has religious significance to the people.

Specifically, the upper edge of the Delta was a zone of contact for different groups for centuries. Elem Kalabari (New Calabar) and Ibani (Bonny), the two principal trading states of the eastern delta, emerged because people migrated there from their original locations in the central delta (Northrup 1978). The founders of Bonny migrated east, bringing them in contact with the Ndoki on the Imo River, they later moved south maintaining strong ties with the Ndoki through trade and intermarriage. There were also high levels of interaction among the Ijo, Edo, Igala and Ibo peoples as far back as the fifteenth century. Northrup shows that cultural complexity characterised the region. Communities made contact, blended cultures and sometimes languages. He argues that immigrant bands that established dynasties were not mass movements of populations or the original settlers on their respective locations. Rather, those that established themselves over indigenous populations were small groups that gained acceptance among their indigenous hosts. Whereas, where peaceful coexistence was impossible, contact led to indigenous migration.

The history of the Delta is not one of distinct and hostile ethnic groups. Over centuries, significant levels of interaction, hostile and peaceful, took place between the numerous peoples of the region. 'Continuous contact between contiguous groups produced broad zones of culturally and sometimes linguistically mixed communities rather than sharply delimited frontiers' (ibid.: 47). The growth of trade was yet another dynamic behind the influx of diverse groups into the Delta. Over time, the major trading communities incorporated immigrants from other communities in the region, resulting in acculturation.

According to Alagoa (1971), strong evidence of early commercial and cultural contacts between Benin and Niger Delta states exist. He delineates three stages in the attempt by Delta states to derive livelihood from their environment. These include an early era of subsistence fishing, hunting and gathering. Exchange with other Delta communities involved in various levels of agriculture in the freshwater zone supplemented subsistence. The second stage of more extensive exchange of produce and long distance trade with the hinterland and finally, trade with Europeans on the coast transformed the Delta states into commercial centres of redistribution: collecting European goods for sale in the hinterland, and receiving hinterland produce for coastal trade (Nwabughuogu 1982).

Writing about two notable intermediaries, Gertzel asserts:

> Nana of Itsekiri and Ja Ja of Opobo, had commercial organizations which stretched over considerable areas of country and which employed several thousand people in various capacities (canoemen, traders, labourers, warriors, local buying agents). No European firm, even if prepared to employ Kru labour on a vast scale, could have done the same.
>
> (Gertzel 1962: 362)

Those who suggest that significant commerce and the wealth of intermediaries like Nana and Ja Ja were the result of European impetus diminish the importance of the extensive pre-European, internal trade (Jones 1963). The next section examines this point further.

Atlantic slave trade and abolition

Although sub-Saharan Africa was unknown to Medieval Europe until the fifteenth century, evidence suggests that in classical times European traders and explorers had contact with West Africa (Crowder 1973: 66). In the fifteenth century, there were two good economic reasons for the Portuguese to start exploring the West African coast, free access to significant gold supply and locating a sea route to India, which would avoid Arab intermediaries. There was also the search for a Christian king in tropical Africa that would join the campaign against Islam (ibid.: 67). By 1480, the Portuguese had explored the West Coast and engaged in gold and pepper trading with Mina and Benin respectively. They settled on a small island, São Tomé and made it the centre of the pepper trade with Benin. Without an indigenous population on the island, Benin soon became a source of labour to the island plantations. The profitability of exporting slaves led to increased demand after 1493. To facilitate trade in humans and other cargo, the Portuguese established a factory, or trading station in Gwato, Benin. Diplomatic exchanges soon followed, culminating in the dispatch of a Benin prince to Portugal and Christian missionaries to Benin. In 1553, Captain Windham was shocked to find that the King of Benin could speak Portuguese.

By 1510, trade was almost exclusively slaves. England soon established itself as the leading trader on the coast and chief exporter of slaves. Conservative estimates show that 24,000,000 slaves were taken from West Africa and Angola, and probably 15,000,000 of them survived the notorious 'Middle Passage' across the Atlantic.

> In the sixteenth century, about 1,000,000 slaves were transported to the Americas, in the seventeenth century, some 3,000,000, and in the eighteenth century, some 7,000,000 or 70,000 a year. Of these, about 22,000 were shipped annually from ports in Nigeria. Benin and its colony of Lagos sent about 4,000 and the ports of Bonny, New Calabar and Old Calabar, which grew up directly in response to European demands for slaves, together with the Cameroons sent some 18,000.
>
> (Crowder 1973: 72)

Crowder argues that the discovery of the Americas and realisation of their mineral and agricultural endowments precipitated traffic in humans. The Spanish policy of settling and developing the New World, the decimation of the populations of the West Indies and mainland and brutal Spanish rule as well as the scarcity of labour from Europe led to using 'Africans who survived well and were adaptable to work in the mines and plantations' (ibid.: 72–73). The Dutch became the leading slavers in the seventeenth century, but lost control to Britain and France after the Treaty of Utrecht in 1713.

Crowder argues that the slave trade had notable impacts on the political structure of the Niger Delta. Before the advent of the Portuguese, the Ijo people who escaped Benin domination migrated to the Niger Delta, living in small, scattered fishing villages. The onset of the slave trade stimulated the growth of these small fishing villages into trading states. The Ijo traded with peoples of the hinterland, exchanging salt and fish for vegetables and iron tools. The slave trade altered this trade pattern. Sparsely populated, the Ijo had to look to the more populous hinterland communities to import slaves, mostly sold to the Europeans on the coast. Yet the Ijo communities retained some of the slaves resulting in a rapid population growth. The Ijo experienced social transformation in other ways (Crowder 1973: 81–82).

The idea that some European traders fostered the emergence of critical State institutions, which led to growth, grossly exaggerates the creative impact of European trade (Alagoa 1970). Niger Delta states had evolved structures before the arrival of the Europeans. The new overseas trade was inserted on the pre-existing long distance trade within the region. Thus, the new trade 'merely altered the nature and dimensions of trade within the Niger Delta, and accelerated changes already begun by the internal long-distance trade' (319). Overseas trade arose at pre-existing locations of authority, requiring no need to fashion new institutions of control. Anene (1966) corroborates Alagoa's position when he argues that the Ibeno of Bonny and the Efik provide the best illustrations of sophisticated political organisation.

The British abolished the slave trade on 1 January 1808. The abolition is a paradox against the backdrop that Britain had become the leading carrier of slaves. The act had profound impacts on Nigeria and marked the onset of legitimate trade, precursor to British colonisation of the country (Apena 1997). The abolition followed 30 years of agitation in Britain against the trade, which spanned more than three centuries. After the abolition of the slave trade, Britain took steps to suppress the trade (Hopkins 1968). This led to a new economic order in the Niger Delta, one that reverted to trade in natural products, mainly palm oil. Growing British interest in the palm oil trade brought a shift in British attitude toward local rulers, including disrespect for their sovereignty (ibid.: 131).

Hinterland trade and the Niger Company

European penetration of the interior led to the breakdown of the monopoly enjoyed by African intermediaries. At the time, the British believed that if trade was to flourish, then the highest authority in the land should be the British government and not African chiefs (Crowder 1973: 150–151). Elements within British commercial and humanitarian circles as well as members of government opposed the idea. Booming trade led to extensive use of slaves as porters and labourers to collect palm nuts and to carry oil to the coast. Thus, rather than substitute trade in slaves, palm oil trade increased the need for slaves. The Delta was unsurpassed in the palm oil trade. In 1855–56, the Delta exported 25,060 tons of oil, over half the quantity of oil exported from Africa (Aghalino 1998: 152).

The policy of Britain in colonial West Africa was to advance British commercial interests (Hopkins 1968). Thus, Britain soon achieved commercial control over the coastal region through the efforts of her consuls and military. Moreover, unequal treaties were deployed to pressure and coerce native governments (Dike 1962). To break the resistance against penetration of the interior, Britain strengthened rulers of the city-state, Old Calabar, who collaborated to advance her interests. Those who refused to support British designs, such as Pepple, King of Bonny, suffered victimisation.

The shift was a function of the desire to capture an expanding hinterland trade (Aghalino 1998). At the time, four British companies were operating in the Niger valley: The West African Company (Manchester); Messrs Alexander Miller Brothers & Co. (Glasgow); The Central African Trading Company (London); and James Pinnock & Co. (Liverpool) as well as numerous small firms and individual merchants (Dike 1962: 204). Together, the companies employed 14 steamers in the Niger trade opening commerce 600 miles into the hinterland. The British government realised that naval power at the coast was of little use to her traders in the hinterland. With the lucrative interior waiting, firms put considerable pressure on the British government. The government rationalised that commercial profit is a function of political security and accepted the view proffered by the traders that trade and political frontiers must expand simultaneously to reach their mutual objective (Hopkins 1968).

Employing armed boats, the companies penetrated the interior, establishing factories at various locations, thus intercepting trades that previously passed through intermediaries, leading to a number of confrontations between Europeans and Africans between 1871 and 1879.

> Military expeditions visited the Niger basin annually, destroying Delta and hinterland towns that had attacked British life and property. So long as the warships remained in the vicinity of the trading posts a thriving trade continued; during the seven months of the dry season, when the ships could not ascend the Niger River, Africans resumed their attacks on the invaders. War and trade alternated with the seasons.
>
> (Dike 1962: 207)

Sustained attacks on British trading posts in Onitsha led to naval bombardment and razing of the town in 1879. Similarly, bombardments occurred at Idah and Abo, just as they destroyed Yamaha, an inland trading station on the Benue River for attacking British traders. The fallout of the forceful penetration of the interior, local resistance and reprisal bombardment was an environment of insecurity that made trade sometimes impossible. By 1879, it became obvious that the peaceful exploitation of the lucrative and substantial trade in the region demanded some form of security.

Concomitantly, from 1877 to 1879, Goldie Taubman unified competing British firms in the region into the United African Company, and successfully eliminated foreign competition. However, Taubman's overriding interest was to bring British political domination over the Niger basin. Taubman was of the strong view that trade and civilisation was impossible in an unsettled interior, and that 'pacification' was, therefore, crucial. In pursuit of his ambition, he established more than 100 trading posts on the Niger and Benue Rivers. He commenced treaty deals with African rulers. By 1884, he had concluded 37 treaties. The figure rose to 237 by 1886. The treaties 'invariably ceded to the National African Company, "the whole of the territories of the signatories", conferring in addition the right to exclude foreigners and to monopolise the trade of the area' (ibid.: 212). Dike (1962) argues that the company had more than 20 gunboats capable of navigating the Niger year round for purposes of policing and pacification.

The people of the Niger Delta opposed and bitterly resisted the company's rule. The company's factories in Akassa, Patani, Brass, Asaba and Idah were attacked. Despite such uprisings, the company's superior firepower kept the locals subdued (ibid.). Dike argues that the motivation behind the eventual annexation of the region was the need to marginalise and neutralise African intermediaries, seen as obstacles to the lucrative trade with the interior. Thus, having successfully claimed the Niger Delta and lower Niger at the Berlin

Conference of 1885, based on the efforts of its consuls, the British, on 5 June 1885, declared a Protectorate over the Niger area.

However, such proclamations were insufficient to transform the area into an effectively occupied territory. The British achieved effective occupation by force (Anene 1966). The British coerced native chiefs into signing treaties they barely understood, by which they ceded their sovereignty, the rights of their people and their lands. However, by West African customary law, it was beyond the competence of the ruler to alienate land. Given that such voluntary surrender of sovereignty was not intended, African chiefs resisted European encroachment. Thus, between 1885 and 1900, British forces embarked on the 'subjugation and pacification' of the Niger Delta (Dike 1962: 218).

Colonial coercion

The colonists utilised force in the establishment, consolidation and expansion of the Southern Protectorate. Tamuno (1978) argues that the protectorate troops and police were employed in patrols, punitive expeditions, or maintained as a threat, and were infamous for their terror. Thus, between 1901 and 1903, there were 13 punitive expeditions and patrols in Southern Nigeria, and such exercises became an annual occurrence. He observes that through punitive expeditions much of the Southern Protectorate became subjugated territory including Bonny, Brass, Degema, Ahoada, Asaba, Calabar, Warri, Opobo, Ogoni, Benin, Eket, Awka and Udi. The natives in these areas had difficulty adjusting to the new colonial government, which in turn sparked public unrest and conflicts.

The British navy gave cover and protection to European traders on the coast dealing with African intermediaries, chiefs and rulers. In the event of crisis, warships were quickly mobilised and deployed, and during war, 'armed cruisers and vessels were generally deployed to blockade the coasts and to support the land forces' (Mbaeyi 1982: 201). Following the abolition of the slave trade, the British navy played an important role in combating the trade by patrolling the coast and seizing slave ships. Trade in palm oil flourished in place of slave trade. The navy once again had the responsibility to protect the substantial volume of trade in the region. Mbaeyi (1982) argues that the unofficial pacts between the navy and merchants, and between the navy and colonial administrators, emerged and remained for much of the nineteenth century.

Conclusion

The physical geographical area known as the Niger Delta was profoundly shaped by economic and political activities between, first various communities in the region and beyond, and second by explorers/traders and colonists. Economic and political processes brought about the trade in human beings and the various conflicts and wars it engendered. Later, trade in commodities led to export reorientation of trade in the region. The booming trade instigated European traders' determination to penetrate the interior of the region. Local community resistance against penetration occasioned several violent conflicts and eventually colonisation. The peoples of the Niger Delta have a rich history predating the arrival of the Europeans. This history is dynamic: a history of peaceful coexistence and violent conflicts; isolation and interaction; and identity and cultural borrowings. That history reflects the changing economic, political and social fortune of Niger Delta communities. Following the forcible penetration of the interior, trading towns like Gwato, Opobo, Calabar, Benin and Aboh lost their early advantage as centres of religious and commercial contact with the Europeans. Things changed with the colonists' decision to establish administrative headquarters in the hinterland. Such relocation effectively sowed the seeds of the spatial stratification pattern that came to dominate relations between Niger Delta communities and the hinterlands.

Despite the eventual violent subjugation of the people, the British colonisers failed to transform Nigeria in pre-determined ways. In the ensuing encounter, however, they did occasion a social formation neither they nor the colonised completely determined. This contact and attempt to engineer diverse peoples in line with European dreams form the foundation from which emerges the root of the crisis unravelling in the Niger Delta today. The historical experiences of communal subjugation and loss of autonomy, attempt at the homogenisation of disparate people, uneven regional development, colonial expropriation of resources, underdevelopment and communal powerlessness continue to reverberate in the present.

Aside from precipitating the insertion of Nigeria into global capitalism, colonialism bequeathed legacies of predation, inter-ethnic competition and rivalry and marginalisation. Although capitalism did not invent exploitation and resistance against exploitation, pre-capitalist formations sustained social equilibrium by balancing peasants' surplus transfers with ruler's provision of security for the cultivators (Wolf 1969: 279). These economic orders changed with the onset of colonisation (Mehretu 1989: 99). Colonial incursion initiated an economy dominated by forced raw materials export, and deliberately reorganised the local economy to respond to European industrial needs. It laid the foundation for an equally intrusive and predatory postcolonial development.

3 Locale
Political and cultural context of mobilisation

Introduction

Locale refers to the formal and informal setting in which everyday social interactions and relations are constituted. In the case at hand, the chapter traces the formal national political space the Ogoni found itself in during the colonial and postcolonial periods. Also, it examines how the Ogoni negotiated their fate within the regional political space in Nigeria, and the institutional contexts of the oil industry and Kagote, the elitist Ogoni cultural group that arrogated to itself power to speak for and on behalf of the Ogoni.

Environmental transformation is mediated by formal and informal interactions (direct and indirect) between translocal actors and local communities. Environmental degradation might create repulsion to local people just as the latter will experience moral shock when interactions with global capital instigate processes that undermine the world as the locals know it. What happens when local worldview and understanding based on an moral ethos come under assault by a capitalist ethos? This chapter helps us navigate the intense pressures modern and new impulses subjected the Ogoni and how the latter tried to make sense of their fate in the new environments. The idea is to lay bare the contextual factors that shaped Ogoni mobilisation against the State and global extractivism.

The national and institutional space

British colonial adventurism welded together hitherto autonomous nations into a plural society or geopolitical entity called Nigeria. The 'contraption' was marked by variations in colonial administrative practices and politics from very early in spite of Nigeria's common colonial experience. A major step in the making of the nation-state was the amalgamation of the southern and northern protectorates in 1914. Despite the unification, little attempt was deployed towards 'making' a nation-state that provided level playing field for all, regardless of provincial or religious sympathies. For over four decades, the colonial authorities operated radically different administrative structures to the east, west, and north of the Niger (Tamuno 1970). Within that period, the system of indirect rule and the attendant policies of preserving the emirates from Christian missionary and Western education influences safely isolated the North and South from each other, but more than that 'it indoctrinated the emirs, chiefs and the emergent nationalist elites in their historical differences, not only political but racial, religious and cultural' (ibid.: 11).

In its enterprise of modernisation, the colonial State initiated processes of social change that served to reorient people to modern systems of societal rewards and paths to such rewards. Through colonial imposition and/or willing acceptance, people's aspirations and expectations changed in the face of the new systems of values. However, the opportunities in the economy and politics were limited. The result is that the various peoples brought together began to compete for positions in commerce, politics and others spheres of life. Those overwhelmed by the insecurities of change sought refuge in 'the communal shelter of 'tribalism'' (Melson and Wolpe 1970: 1115).

The dominant element of Nigeria's federalism was the prevalence of one ethnic group in each region: Hausa–Fulani in the Northern Region, the Yoruba in the Western Region and the Igbo in the Eastern Region. Thus, these three became the power brokers in Nigeria. Nigeria as a concept came to represent a tripod, a land of three ethnic nationalities, rather than a polyglot, a land of many ethnic nationalities (Omoruyi 2000). This disjunction in the process of social change became the root of communal consciousness or ethnicity as a social force in Nigeria: the fears as well as the facts of predominance of the members of one administrative or political region, one community, one locality or one group of individuals over others in the struggle for opportunities, resources and power. The process did not change after independence. In fact, it has been extended and deepened (ibid.: 15).

To the regional elites, their fortune was tied to their communal origin or regions. Moreover, the fortune of their region was seen as threatened by competition from other regions. As a result, elites began to organise and mobilise communal origins and ethnic associations. Ordinary folk in turn perceive the political elites as champions of their group interests and success of the latter as the success of the ethnic/regional group in its competition with others. The mobilisation of one group sent threat signal to other groups, and mutual suspicion of threats to group aspirations developed. Given differential rates of social mobilisation, first mover advantage placed the groups that mobilised later at a disadvantage in terms of access to power and wealth.

The isolation of the Islamic northern protectorates from Christian missionary activities and Western education gave the south a head start, which reflected in its domination of administrative and technical jobs all over the country. Other factors that shaped differential mobilisation were variations in timing of contact with the Europeans, and regionalisation of politics. The consequence of differential mobilisation is that new 'socio-economic categories therefore tend to coincide with, rather than intersect, communal boundaries, with the consequence that the modern status system comes to be organised along communal lines' (p. 1116). Within each region, differential rates of mobilisation were at play widening cleavages between the dominant ethnic group and minority ethnic groups in the region. In the Niger Delta, for instance, differential rates of contact and access to the Europeans and economic growth widened the latent cleavages between the coastal communities (Ijaw) and their interior neighbours (Ogoni).

To further their political aspirations and interests, the western and northern region established ethnic associations, mirroring the eastern region. They established largely regional political parties in opposition to the National Council of Nigeria and the Cameroons (NCNC) set up by Nnamdi Azikiwe that became strongly seen as an Igbo party. The proliferation of ethnic association served to provide a fertile context for the exacerbation of competition among the three dominant ethnic groups. Following examples at the national level, minorities within the regions also established their own ethnic associations and political parties to pursue their own aspirations. Concerned with the advancement and welfare of their communities, such ethnic associations provided public services including scholarships, education, health, etc. and thereby promoted ethnic loyalty. For instance, the Ibos and Ibibios seized the initiative, organising communal associations as early as the 1920s and 1930s. Their example inspired other ethnic groups across the country to set up their own ethnic associations (p. 1120). At the regional level, ethnic associations such as the Omo Egbe Oduduwa (Western region) and Northern People's Congress (Northern region) served as ethnic pressure groups, and deepened communal consciousness and promoted communal interests within the respective party and government.

In the ambience of bitter rivalry, nationalist politicians failed to take advantage of the opportunities to make a constitution for socioeconomic engineering. Instead, it became a field for squabbles, and acquisition of political power. The question of erecting the fundamentals of a stable and viable postcolonial state receded from view.

> The North saw in the unfolding political landscape a topography of southern domination of the administrative cadre and the economy because of their sociohistoric privileges and advantages of Western education. The South, on the other hand, could sense a deadly collaboration between the receding colonialism that British officials represented and the steadfast conservatives of the northern emirates to capture political power at the centre to the detriment of the more educated south. Between the two southern regions there was no love lost; keen competition for employment and other economic resources continually fouled their mutual relationships, perceptions, images, and attitudes.
>
> (Ajayi and Ekoko 1988: 252–253)

Capturing power was the heart of politics in colonial Nigeria; there was little interest in majority rule and political equality. Despite seeming differences, the colonisers and nationalists were divided over the question of who should rule; not the issue of governance (Ake 1973: 358). The nationalists demanded a share of power, rather than reform of colonial authoritarianism. Colonial politics came to signify a mortal struggle for capturing power. As political competition intensified, political actors engaged in whatever would realise their objectives rather than what was legitimate or legal. Following British colonial example, the ruling elites turned the State, the source of economic surplus derived from rising oil revenue, into an instrument of private capital accumulation (Schatz 1984). The result of intense political competition was 'political anxiety' or fear of the backlash of not being in control of State power (Ake 1973: 357). Such perception of politics continues to shape the contemporary view of politics as strictly a struggle for rulership (ibid.: 359).

In the passionate competition for power, the overriding consideration was who should control the distribution of scarce federal resources and patronage and not developmental policies. The 'post-colonial state became a forum for trading power to protect largely regional and local interests, often to the neglect of the overall national interests' (Ajayi and Ekoko 1988: 267). The primacy of regional interest over national benefits continues to influence economic decisions. In the First Republic, development planning was an evocation of regional bias rather than national development. The ruling political elites preferred projects that benefitted their regions to cheaper economically viable alternatives with national benefits. Despite clear evidence that modest investment in natural gas, abundant in the south, would generate enough electric power supply, the NPC government favoured bigger investments in the construction of the Kainji Dam. Similarly, the government was not keen on the idea of exploiting iron ore located in the north to develop a steel industry in the south (ibid.: 268).

The question of erecting the fundamentals of a stable and viable post-colonial state receded from view. In consequence, institutions of the postcolonial state failed to deliver the expected benefits and instead became tools of control and exclusion from political participation. The prevailing politics is 'a fight to capture and privatize an enormous power resource' (Ake 1996: 8). It is far from a fair competition to elect those who would represent the collective will. As a result, Ake asserts, there is no state, or a public space, that embodies a collective identity. There is only a contested terrain where interest groups and communities engage in a battle for appropriation. The 'act of doing battle constitutes us as a purely negative unity' and as a 'polity of takers rather than givers' (ibid.: 8). And in that polity, the ruling elites hanker after state power to appropriate as private property (ibid.: 8).

The unending struggle between the three major groups to control the federal state, the non-inclusion of the minority groups in the final constitutional document signed in London, the regionalisation of politics, which gave the impression that Nigeria was a natural tripod, which effectively wrote out the minorities as well as the domination of the minorities within each region by the regional hegemony contributed to minority ethnic identity construction.

The regional institutional space

The Ijaw refers to a group of some 8–12 million people, segmented across the six states of Nigeria's Atlantic coastline. They are arguably Nigeria's 'fourth largest' ethnic group and, although a minority in the Nigerian context, the largest minority group in Nigeria's oil-producing Niger Delta region. The Ijaw have not always been members of a single ethnic identity. Emergent Ijaw nationalism claims the existence of a relatively homogeneous group. History, however, suggests that pan-Ijaw ethno-political consciousness is a relatively recent development. Competition with the Igbo majority of Nigeria's Eastern Region has served as an important crucible for the emergence of an Ijaw nationalist identity.

Early exposure to Europeans since the sixteenth century, through the intermediary roles played by eastern and central coastal Ijaw as 'middlemen' in the trans-Atlantic trade in mainly hinterland Igbo slaves and later in palm oil products, created complex social and political institutions and strong 'city state' rather than specifically 'Ijaw' identities. These were particular to the coastal Ijaw whose creeks and rivers were mainly saltwater ones bordering the Atlantic coast in the central and eastern Delta (Nembe, Kalabari, Okrika, Ibani). This early 'exposure' to all things European paved the way for the 'warrant chief'/'court clerk' role played by many of their elites in the early colonial administration.

By contrast, many upland freshwater Ijaw communities in the northern parts of the central and western Delta had more in common with Igbo societies further north as sites of slave raids and later of palm oil production. This is supported by Alagoa (1986) and Okorobia (1999) who argue that some slaves and items sold by the Ijaw came from within Ijawland. In addition to being geographically closer to the Igbo hinterlanders, the freshwater Ijaw shared similarities in social and political structures leaving them open to caricatures of 'backwardness' in the eyes of Ijaw 'kin' from the more 'civilised' coast (Jones 1963).

The seeds of a pan-ethnic political identity, however, were gradually sown in the early 1930s and 1940s amid growing competition for influence and resources in the colonial administrative centre of Port Harcourt. Ijaw migrants and other Delta communities found themselves in competition for jobs, political influence and land with the vast numbers of Igbo migrants from the north of the Niger Delta.

In the context of decolonisation, the creation of three regional political units in 1946, each with an ethnically dominant 'majority' group and a large number of smaller 'minorities', became the crucible of trans-ethnic 'Rivers'/ 'minority' identities developed to do battle with the hinterland Igbo. Throughout the 1950s period of episodic deliberations over Nigeria's constitutional future, a Calabar-Ogoja-and-Rivers political identity also emerged, which included an even larger stretch of non-Igbo minority groups living in three provinces in the eastern region. Both minority political platforms used notions of cultural similarity to make a case for greater 'minority' autonomy vis-à-vis the politically and numerically dominant Igbo ethnic group that dominated the Eastern Region and controlled the largest political party, the NCNC, which also controlled the Regional government (Tamuno 1972).

Many had long suspected the Ijaw of using 'Rivers' as a smokescreen for what was effectively an exclusively Ijaw political project. Not all Ijaw, however, supported the idea of creating a Rivers state, particularly the politically and economically powerful Kalabari chiefs who advocated a COR state (based on the boundaries of Calabar, Ogoja and Rivers Provinces) rather than a Rivers state per se. New rivalries emerged within the Rivers coalition. The Rivers Chiefs and Peoples Conference, led by Harold Dappa Biriye, who had championed the creation of the first Ijaw ethnic associations in Port Harcourt, and led the Rivers state creation cause, split into two factions. The larger faction, made up of chiefs and some younger educated men, opted for accommodation with the NCNC government. A smaller faction organised around a new party, the Niger Delta Congress (NDC), created by Dappa Biriye in May 1959, was opposed to any accommodation with the Eastern Region's NCNC government. The NDC formed an alliance with the rival party in the Northern Region, the NPC (Northern People's Congress) prior to the 1959 Federal elections, ostensibly because the NPC 'promised', once in power, to create a Rivers state as a Federal territory.

Encountering institutional spaces

Unlike the Ijaw, by 1932, well over two decades into colonial rule, a pan-Ogoni consciousness did not exist largely because of communication difficulties among and between the Ogoni cultural zones. Ogoni territory was balkanised into two administrative divisions; Eleme was grouped in the Owerri Province, and the other Ogoni cultural groups grouped in Calabar Province. This prevented political contact between its peoples and other groups, and, in turn, prevented a pan-Ogoni consciousness (Isumonah, 2004).

In 1932, the Eleme requested that they be placed in the same administrative unit as the rest of the Ogoni. However, few Ogoni showed interest in the idea of constituting the entire Ogoni into an administrative area. Partly because of the coordinating role of the Ogoni Central Union (OCU), which was formed in 1945, the Ogoni administrative area, or Ogoni Native Authority, came into being in 1947. During constitutional reforms in the 1950s, Ogoni foremost nationalist prioritised the issue of Ogoni marginalisation and demanded more social and economic opportunities for the Ogoni through scholarships, development projects and political appointments from the then Eastern Regional Government as measures to undo Ogoni marginalisation.

Memories of experiences garnered in the course of contacts and varied interactions in common institutional settings and processes organised by the colonial government deepened the critical need for an Ogoni identity. For instance, there is evidence that Ogoni experienced unpleasant and unprovoked ethnic-based insults at the hands of their Igbo compatriots. Such experiences generated resentment, which forced victims toward ingroup solidarity. Ogoni experiences furthered identity formation in two ways; the experience of external labelling that demarcates the victim as different, and the consequent resentment that drives victims to explore and demarcate their or distinctiveness or identity.

Minority groups in the three regions that composed Nigeria, including the Ogoni and Ijaw, expressed fear of ethnic domination in independent Nigeria controlled by elites of the three dominant ethnic groups. The Willink Commission of Enquiry was created in 1957 to determine whether ethnic minorities' fears were well founded. The Ogoni demand for creation of a state comprising them and the Ijaw out of the Eastern Region met with opprobrium from ordinary Igbos who jeered derisively at Ogoni traders in the predominantly Igbo towns of Aba and Port Harcourt. In reaction to these experiences, the Ogoni managed a protest vote against the NCNC in 1957, which led to the Action Group (AG) winning the two Ogoni seats in the Eastern Region's House of Assembly.

In 1950, three decades after the Ibo, Ijaw and Ibibio headstart, Ogoni elites formed the all-inclusive Ogoni State Representative Assembly (OSTRA) as a means to further Ogoni aspirations and interests. OSTRA helped to launch some Ogoni indigenes into state and federal decision-making structures. Thus in 1951, its President, T.N. Paul Birabi, and Secretary, F.M.A. Saro-Wiwa, were elected to the Eastern Region's House of Assembly on the platform of the NCNC. It was not long after that T.N. Paul Birabi left for the Federal House of Representatives on whose platform he participated in the 1953 London Constitutional Conference (Loolo 1981). Presuming an existing Ogoni identity, the Ogoni Divisional Union (ODU), formed in 1962 from the ashes of OSTRA, became the medium for fostering Ogoni interest in the creation of a Rivers state, until this was realised in 1967.

With the onset of the Nigerian civil war in 1966, the Niger Delta minorities did not want to be part of Biafra Republic proposed by the Igbo secessionists. The Ogoni were divided on the issue. While some Ogoni elites joined Biafra, a few led by Saro-Wiwa joined the federal side. The federalist felt that their fate would be worse in Biafra. Okonta (2008) claims that Saro-Wiwa broke ranks with his fellow Ogoni and joined the federal side because he was searching for a place in the sun. However, a more plausible explanation for the choices made regarding the war should reflect some continuity with sociopolitical circumstances of the actor in question. That Saro-Wiwa would find a space to shine on the federal side was not pre-determined. The war could have gone either way. It is a more persuasive argument that individual support for Biafra was shaped by past experiences rather than ethno-political loyalties.

Even if pecuniary incentives shaped an actor's decision, such a decision would have been mediated by prior experiences, and in this case, the possibility of Ogoni prosperity under Biafra. For instance, while most people in the village of Emeyal, an Ijaw community, were opposed to Biafra, Emeyal elites who had been very pro-NCNC supported Biafra. Similarly, the Epie and Atissa youth supported the Biafran army offered them an opportunity escape imposed categorisation and marginalisation by neighbouring Ijaw communities who considered them as not 'proper' Ijaw. The definition of the Epie and Atissa as not proper Ijaw was an imposed categorisation that did not sit quite well with the people. There was need for the Epie and Atissa people to define who they were, and they needed an enabling environment to do so. What materialist explanations ignore is that late-comers to the game of identity formation were manoeuvring in a context over-determined by the dominant groups to create a cultural space for them. Taking sides with the more benevolent provided space for the unhindered articulation of the much desired identity. For some, Biafra provided that space, and for others the federal government offered a better prospect. A materialist calculation of cost–benefit that is insensitive to context does not and cannot explain the decision of Atissa to join Biafra just as it does not explain the decisions of Ogoni elites.

On the eve of the civil war, 5 May 1967, head of state General Gowon took advantage of the distrust to create 12 states out of the existing three regions that composed Nigeria. He carved out two states (South-Eastern, and Rivers) for the minorities from the Eastern region to further undermine whatever support Biafra enjoyed amongst the minorities. The minorities saw Gowon's action as a fulfilment of their age-old desire for autonomy and as the destruction of the foundation of Eastern regional domination. The minorities imagined that their freedom from the hegemony of majority domination would exclude the latter from access and control over their oil wealth. They thought that as a result they would enjoy control over the new oil wealth, and be able to use it in ways that meet their own developmental needs. The minorities' hope, however, proved to be ill-informed (Mbeke-Ekanem 2000).

The state-creation exercise of 1967 gave the Eastern Region minorities their much sought after autonomy and freedom from Igbo domination. In Rivers, however, the Ogoni remained a minority group where the more preponderant Ijaw became the new dominant ethnic group. The possible implication of the new situation was not lost on the Ogoni ethnic association, the ODU. Through its existence and activities the ODU sustained consciousness of Ogoni minority status in the State. In 1968, the Lagos arm of ODU addressed a letter to the Rivers state government reiterating the hardship the Ogoni endured under Igbo domination. The unpleasant experiences informed Ogoni collaboration with other Rivers minorities for the struggle for creation of the Rivers state. Too aware of that recent past, the Ogoni people 'do not want to be treated or regarded as second-class citizens'. It emphasised the need to consider the Ogoni Division as an indivisible administrative unit in the state. Moreover, they advocated that personnel sent to Ogoni Division be of Ogoni Divisional origin not only to obviate communication problems but also to ensure that administrative tools do not become tools for oppressing the Ogoni as was the case in the past.

The hope that accompanied the creation of the Rivers state and the belief that the state's oil revenues would be used for its development quickly evaporated. The oil-based prosperity of the minorities was short lived as the Hausa and Yoruba majority ethnic groups could not stand by and watch the relative prosperity of the minorities. The Gowon and Murtala-Obasanjo governments progressively whittled down the long-standing principle of revenue allocation by regional derivation. Instead, they devised a means of distribution based on population and interstate equality. In the late 1980s, revenue allocation reaching the oil producing minorities dwindled to about 1.5 per cent making it impossible to provide welfare services to the people. Widespread discontent in the Niger Delta deepened. In response, the Babangida regime increased revenue allocation to oil-producing states from 1.5 to 3 per cent and established the Oil Mineral Producing Area Development Commission (OMPADEC) to administer the funds and undertake development projects for the region. The regime also increased statutory allocation from 1 to 2 per cent for amelioration of environmental degradation. However, such redistributive concessions fell below the expectations of the minorities (Mbeke-Ekanem 2000).

Yet, when the Babangida regime created 11 additional states in August 1991, the agitations of the Ogoni and other communities in the Rivers state were completely ignored. Saro-Wiwa interpreted the process and act of state and local government creation as 'robbery with violence' (Saro-Wiwa 1995: 100). It was a strategy designed to transfer the resources of the oil-bearing Niger Delta to the ethnic majorities, given that most of the new states and local governments were in the regions controlled by the dominant groups. Worse still, Saro-Wiwa laments, none of the new entities was viable as they depended on oil revenues for survival. The Babangida regime's failure to entertain calls for creation of additional states from Rivers signalled its disdain for the minorities; 'it was this act alone which so outraged me that I decided that, come life or death, the brutalisation of the peoples in the oil-bearing delta of the Niger would have to be questioned, exposed and brought to a stop' (1995: 99–100). Saro-Wiwa's protestations in academic and literary forms 'had fallen on deaf ears. Something else had to be done to bring the urgency of the matter home' (ibid.: 100).

Shell

It is evident that from earliest times, basic trust was absent in the relationship between the Ogoni and Shell. The lack of trust may have triggered other negative emotions such as suspicion of Shell's intentions, resulting hostility and blame. The discursive conflict between the Ogoni and Shell that followed the events of 9 June 1970 is the clearest indication of lack of trust. Another was the refusal of Shell to dialogue with the Ogoni when the latter issued an ultimatum in 1990. Afterwards, a long series of events stoked Ogoni hostilities. These include the support, material and otherwise, Shell provides to the State in its violent repression of the Ogoni; Shell's refusal to prevail on the State and spare Saro-Wiwa life; the various attempts by Shell to re-enter Ogoni without proper resolution of its problems with the Ogoni; and the failure of a government-organised reconciliation process (Kpalap 2008, interview).

We can draw very useful lessons from the nature of the Ogoni–Shell relationship by examining Shell–BP's response to the Ogoni petition referenced above (Saro-Wiwa 1992: 50–56). From the response, reference NO. PUB/2110, 9 June 1970, it appears that the Ogoni petition was only one in a series of petitions from the Ogoni years earlier. It made clear that Shell–BP's role was to 'find and produce hydrocarbons' and by so doing contribute to national development. The Company argues that development responsibilities was outside the remit of its contract with the government.

Given that the company strives to ensure that its operations cause minimal disturbance, the Ogoni charge that the company's operations exacerbate land scarcity and pay insufficient compensation for environmental damages are inaccurate (ibid.: 50–51). Shell found it worthwhile to persuade any who would listen that the 15 per cent crude oil produced in Ogoni constitutes a small portion of total crude oil produced in Nigeria. The letter impugns Ogoni claims of population density, claiming that the actual density is less than one person per acre based on the 1963 national census, which showed that the Ogoni population averages 564 persons per square mile. Shell-BP claims that the approximate land area of Ogoni is 264,320 acres out of which the company occupies less than 1,000 acres of land for its operations. Thus, Shell denies the charge that its operations have disrupted the entire Ogoni loal economy.

At about this same time, a major incident occurred at the Bomu oilfield. Initially, the company tried to cover the incident up, which meant it did not receive the attention it deserved. It took Shell several weeks to contain the situation but not before extensive environmental damage. Dere community youths bemoan the colonising mode of naming land, which Shell employed arbitrarily, albeit strategically, renaming Dere, Bomu. This is clearly an act of power over the land and its occupants because the so-called Bomu is not an empty or unoccupied space. What such naming does is erase the agency and meaningful relationship between the Ogoni and the environment. The arbitrary and colonially informed practice of naming and classifying people and places aims at control (Adams 2003: 24).

Nineteen years after the Bomu blowout, some families, victims of the blow-out, sued Shell for compensation in Shell v. Farah. In March 1988, Shell claimed it had rehabilitated and handed the land over to the plaintiffs and that it had paid compensation for damaged trees, crops and other elements as well as another for land degradation. The plaintiffs claimed ignorance of Shell claims and consequently initiated legal action in 1989. In court, Shell relied on an expert witness who argued that oil spills were not hazardous but functional to agricultural enterprise (Frynas 2000: 168). The plaintiffs argued otherwise, prompting the judge to appoint referees to reconsider the evidence. Their joint finding corroborated the plaintiff's argument. The judge awarded the plaintiffs Naira 4,621,307 in compensation. Shell appealed the judgment. The appeal failed before the Court of Appeals. In several instances, absurd expert claims helped Shell avoid responsibility for environmental damage (Frynas 2000: 169).

From 1981–1986, Shell had 24 compensation claims lodged against it in Nigerian courts, and in 1998, the number jumped to more than 500 cases (Frynas 2000). The rise in the number of lawsuits suggests that compensation arrived at through negotiation and mediation was unsatisfactory, which indicated that negotiation and mediation are inadequate methods for resolving community compensation claims. Frynas is of the view that given such a gap and the absence of compensation-monitoring mechanisms, victims of oil operations may resort to violence or lawsuits. The latter is more likely if the victim has the means to initiate a legal challenge. Shell accuses the Ogoni of being saboteurs, responsible for 69 per cent of oil spills in Ogoni between1989 and 1994 (ibid.). Shell employs the guise of sabotage to escape liability for damage. Yet, the company has never prosecuted anyone on the charge of sabotage.

Common experiences of exploitative Shell–Ogoni relations provided fodder for the construction of an Ogoni identity by the Movement for the Survival of Ogoni People (MOSOP).

Kagote: Ogoni local elites

Colonial rule gave rise to a new category of privileged individuals, warrant chiefs, who came to exercise tremendous influence in the local community. The position of warrant chief was created by the colonial government, and individuals were arbitrarily appointed into the position with a mandate to execute certain tasks for the colonial government. Therefore, warrant chiefs were not autonomous traditional leaders of their communities. In Ogoni, warrant chiefs and other local elites became the fulcrum of political leadership. The scion of colonial institutions naturally became the foundation members of the Ogoni Central Union in the 1930s and 1940s. The OCU acted as a political representative for the Ogoni and focused its energy on the socioeconomic gap between Ogoni and its neighbours. In the 1950s, the newly formed OSTRA championed the political advancements of Ogoni elites (Loolo 1981). OSTRA quickly disintegrated following intra-elites' conflict and competition for juicy political positions. In the early 1960s, a new organisation was formed to champion the Ogoni cause, including the creation of a Rivers state.

Nevertheless, intra-elites' conflicts continued until the Civil War in 1967. Despite perceived marginalisation under the Eastern Regional Government, Ogoni elites were divided on whether to ally with the federal government or the rebel Biafra government. Saro-Wiwa and a few others supported the federal government. G.B. Leton, I.S. Kogbara and many other elites joined the Biafran side. Following the end of the war, Ogoni elites became direct beneficiaries of the newly created Rivers state in terms of appointment, and wealth accumulation. Basking in the euphoria of freedom from Igbo marginalisation and accompanying social and economic opportunities, Ogoni political elites formed Kagote. Kagote was an elitist and exclusive social cultural organisation aimed at safe-guarding members' interests in the new state.

Nekabari Johnson Nna argues that members of Kagote used their position in government and business to create opportunities for private accumulation and the benefit of organisation members. Given their new wealth and rising political influence, the elites easily consolidated their domination over Ogoni political affairs. Kagote members who were also members of the Rivers State Executive Council mediated the allocation of six first-class chieftaincy stools, the highest figure received by any ethnic group, to Ogoni. Also, Kagote decided who occupied the stools, and who assumed juicy positions in the local government councils and top positions in the Rivers state civil service. Kagote employed its influence to ensure positive outcomes for members in the latter's quest for juicy contracts with the State and multinational oil corporations. Also, Kagote ensured that its members stayed above the law even where such members had breached the law outright. Members of Kagote used State security forces to intimidate and harass ordinary Ogoni with whom they had any differences.

The new men of power in Ogoni were thrown up by the colonial establishment to do the bidding of the latter. Following the end of colonial rule they became the local power brokers. Through successive organisations that were elitist and exclusive, the elites sought to promote the interest of Ogoni vis-à-vis the context of inter-ethnic relations in Nigeria. By means of organisations and clientelistic ties with the state and multinational oil corporations the elites gained access to juicy political positions and wealth accumulation. Thus, their status and power became dependent on such clientelistic networks. Elites' own interests prospered as the State's interest in global capital accumulation prospered in Ogoni. Thus, as it were in the colonial era when warrant chiefs had to do the bidding of the colonial state in order to retain their new found privileges and position, Ogoni elites safe-guarded the extractive industry in Ogoni in order to retain their power, affluence and clientelistic ties to the State and Shell. Ogoni elites could not, and were not, in a position to defend the interests of ordinary Ogoni.

In the unfolding development, youths and women had no place in Kagote, which purportedly represented the interest of all Ogoni. The situation could be explained away as the result of the commonplace understanding of Ogoni as a gerontocratic society where the elders spoke for the society. Youth and women were to be seen, not heard. Granted that Ogoni was led by the elders, the extent to which youths and women were invisible is, however, contestable. Evidence suggests that pre-colonial Ogoni had developed an inclusive system of self-organisation, Yaa, which contributed immensely to the survival and viability of Ogoni. Tonwe-Kpone argues that through the Yaa tradition, the Ogoni ensured their survival. The Yaa was an inclusive cultural system through which young men and women were trained for adult life and recruited into leadership position. The scholar emphasises that through such means, young men and women, and adults participated in the affairs of the community. This society, despite the sparse details, was a world apart from the postcolonial Ogoni society where the Yaa tradition had become moribund, and local elites, constituted into successive social organisations, spoke and acted for the Ogoni. Indeed, as shown above, local elites furthered their own interests at the expense of the ordinary Ogoni.

On the face of it, the system of local elites' domination and exclusion in Ogoni seemed unquestioned and uncontested. Evidence, however, suggests that beneath the calm surface were smouldering grievances. Saro-Wiwa claims that following his tour of Ogoni, he met youths who were angry against a society that disinherited them. Ogoni youths were bitter against the society, which they held responsible for their superfluousness. Disillusioned with the educated local leaders, Ogoni youths came to perceive the former as men who drink beer while the underclass drinks muddy water. The youth further accused local elites of taking advantage of young men's wives because they had money to jail their husbands, in the event of protest. The same elites were those whose overriding interests lie in government positions and Shell contracts. The argument here is that the Ogoni movement could not have been taken seriously if it left domestic power relations unchallenged while crusading for equality at the national level.

Conclusion

The Ogoni emerged from a precolonial era, marked by communal autonomy and egalitarianism to colonial and postcolonial institutional spaces where it had to struggle for autonomy. Lumped together with numerous other ethnic groups to form a new postcolonial state, Nigeria, the Ogoni had to equally struggle to ensure against ethnic domination by the new dominant ethnic groups, regionally and nationally.

The discovery of crude oil in Ogoni brought a sigh of relief but the hope in oil was soon dashed as oil exploitation brought about massive environmental degradation. Ogoni appeals to the government and Shell fell on deaf ears. Kagote, the cultural group that supposedly represented the Ogoni, turned a blind eye to the environmental plight of the people. Ogoni youths saw members of Kagote as local oppressors and devious collaborators with Shell and the State.

4   Landscape, capital and violence

Introduction

The history of the nexus between capital and the Niger Delta has been one of alteration and production of the socio-physical topography of the region. The colonial project of export promotion and arrival of oil firms led to the transformation of socio-physical space and the emergence of a new socio-environmental landscape. The forceful engineering of a shift from food to export crop production among peasants is a case in point. Swyngedouw (1999: 461) remarks on the general conflictual nature of the process of landscape transformation when he posits that the resultant socio-physical landscape instantiates 'historical-geographical struggles and social power geometries'. Thus, existing socio-natural conditions are results of the transformations of pre-existing formations, which are intrinsically social and natural.

The development project consists simultaneously of transformations, which are economic, ecological and cultural in nature (Escobar 2003). Mega-development projects, such as pipelines, petrochemical plants, roads and ports are inherently displacing. Landscape transformation implies a ruthless attempt to destroy the cultural, ecological and cultural differences intrinsic to a place and embodied in local practices. Mega-projects, supported by the State's ideology of modernisation and economic growth, with sophisticated technology and heavy equipment, overwhelmingly aid such transformation of the landscape.

Oil development in the Niger Delta is a geographical project embodied by intense spatial transformation. The alteration of nature and society, thus, reflects the inherent contradictions of development. Against the project of a nature–society dualism, it is critical to reflect that 'community and environment constitute a single, integral and open system; they are mutually responsive to, reciprocally constructed and informed by, one another' (Whitt and Slack 1994: 24–25). From such a decolonised framework or hybrid perspective, land enclosure or theft is not merely a disruption external to a local community with resultant negative impacts. Land is an integral part of such a community. In other words, violence to nature is inherently violence to the social fabric.

What makes oil-induced displacement of significance is the fact that it happens in a place; a portion of space invested with meaning by a group of people and to which they have become attached. As Cresswell (2004: 11) argues, place is a way of seeing, knowing and understanding the world. A non-place approach to the Niger Delta sees nothing but a world of vast oil resources and nature. However, the conception of the region as a world of places enables a deeper perception of worlds of meanings and experiences, connections and attachment between people and places. A place perspective frees us from the rationalising gaze of modernity, forcing on us a decolonised understanding that the so-called nature, far from being pure, is the outcome of both social and natural processes. A worldview founded on the modernist understanding of nature as unspoiled and as space, empty and without meaning, not only does violence to constructed nature, it vaporises the basic coordinates by which people negotiate life.

As an ongoing process of change, oil development embeds in a series of power relations legitimated through the contested discourses of progress, unity and development. In experiential terms, however, development remains a penalising phenomenon for most communities in the region, generating contradictions, the effects of which are marginalisation, human misery and the emergence of the superfluous. It is within this development context that inversionary discourses that threaten the order of things and challenge the modernist logic of environment emerge. This chapter examines the basis and processes by which such discourses arise or why and how the transformational impacts of development on humans and non-humans come to meet with resistance.

Views of development

In the 1980s, it became apparent that the various paradigms of modernisation, including the import substitution industrialisation strategy and 'Operation Feed the Nation', failed to boost meaningful development in Nigeria. This awareness provoked forceful criticism of development economists by their neoliberal counterparts. The latter blamed internal factors for Africa's economic crisis. The Berg report blames domestic policy for the crisis in Sub-Saharan Africa (World Bank 1981). However, the report was silent on the role of external actors in shaping domestic development policies in Africa. The neoliberals held that state intervention obstructed the operation of free market forces and efficient allocation of resources, resulting in patrimonialism and corruption (Dibua 2006). Neoliberal theory, thus, reduces people to isolated creatures of the marketplace, devoid of history, culture and environment. It disregards power relations and social and historical contexts of state intervention, and despises historically formed meanings and values (Brohman 1995: 297–305).

Whether apprehended as state or market-led, development theory and praxis mean different things to different people. To some, it represents immanent or intentional development (Cowen and Shenton 1996), planned public, private or combined mobilisation of resources in the promotion of economic growth (Leftwich 2000: 22), or an unending process of economic growth. One may also view development as the expansion of freedoms (Sen 1999) or as a discourse of domination (Leftwich 2000: 63–68). For Escobar (1984: 384), development has not only failed, it remains a discourse or tool by which Western developed countries create the Third World and seek to manage and control it. It is 'a series of political technologies intended to manage and give shape to the reality of the Third World'. Moreover, given that representations do not reflect reality but constitute it, development discourse constitutes the problem it seeks to analyse and resolve (Escobar 1995: 130).

Escobar's conceptualisation of development as discourse suggests that there exists a single encompassing development discourse. Such a view neglects alternative and competing discourses such as basic needs and development as freedom approaches. It creates a dualism of an impervious and top-down development discourse and a bottom-up anti-development discourse, leaving little space for middling discourses that allow for heterogeneity, exchange of experiences, ideas and responsiveness to local views (Grillo 1997: 24–25). It becomes difficult to explore the varieties of struggles and alternatives at the grassroots that do not conform to such dualism. Grillo upholds the idea that development consists of multiple voices and sets of knowledge even if some voices are more influential. A view of development as composed of multiple voices and practices, rather than a single hegemonic discourse, enables an examination of the complex and contradictory relations between development discourses, and facilitates an understanding of the heterogeneous and conflicting strands of thoughts within particular discourse.

Development is irreducible to discourse because there 'is no materiality that is not mediated by discourse, as there is no discourse that is unrelated to materialities' (Escobar 1995: 130). This implies that materialities have their own existence independent of discourse and that reality affects discourse when it responds to changing situations (Parfitt 2002: 30). Escobar (1995: 46, 145) admits that the work of development agencies does contribute to the amelioration of practical human problems such as poverty. Cowen and Shenton (1996: 454–455) emphasise that development is not only composed of doctrines, but also by 'the practice of development'. The processes of development always involve 'the organization, mobilization, combination, use and distribution of resources *in new ways*' that inevitably result in disputes over how the resources are to be used and who should lose or gain (Leftwich 2000: 5). Approaching development as a set of conscious action geared at a desired goal is beneficial to this task.

Development as oil extraction

Traditional common sense construed primary commodity production for export as a primary engine of economic growth (North 1955; Mikesell et al.

1971: 16). Some argued that given the comparative advantage in the production of primary goods, developing countries should allocate a substantial portion of their productive factors to raw materials production and exports (Mikesell et al. 1971: 16). Critics argue that the benefits of trade in primary commodities accrue to industrial countries and that concentration on raw materials export could hinder industrial growth. Other economists stress, 'the role of resource industries as a leading sector that, under certain conditions, can induce broadly based development' (Mikesell et al. 1971: 17; Hulme and Turner 1990: 101). Although the governments of newly independent countries showed diffidence toward transnational capital, seeing them as neo-colonial agents (Koenig-Archibugi 2004: 16), the prevailing belief was that foreign direct investment (FDI) represented a *sine qua non* for the economic transformation of developing countries (Koenig-Archibugi 2004: 241; UNCTAD 2002).

The arguments levelled against export-oriented development are organised around three points. First, the creation of export enclaves that directs earnings to importation of consumer goods and non-productive investments, an overvalued exchange rate and the attraction of skilled labour and capital from the rural areas by the booming sector, leaving the non-export sector poorer, and uneven regional development. Second, fluctuations in global demand and prices for primary commodities, decline in prices relative to manufacturers. Third, the need for sustainable mineral exploitation and use. FDI relates to higher levels of conflicts and regime instability, with a possible weakening effect on the ability of state actors to design and implement workable development policies (ibid.: 140). Mikesell et al. press the point that mineral dependence and the private capture of public mineral wealth is not a valid reason against such development (1971: 20). It is hard to prove that exporters would have been better off without mineral exports or that the presence of the latter does not currently contribute to their potential for development if they adopt appropriate policies (ibid.: 20).

A teleological assumption of an inevitable link between investment in the extractive industry and economic development was characteristic of (neo) Marxist development theories and the modernisation theory (Schuurman 1993: 12–13). In both sets of theories, the tendency was to relate the entire process of planning, action and effects, and to assume that the three stages were completely within the control of human intention or agency (Ferguson 1994). It failed to register that the outcomes of calculated human activities or development can spin out of control (Elias 1991: 62). Such modes of thought remain prevalent and largely inform state-led or neoliberal development and, why they fail (Scott 1998: 3–5).

Given the growing divergence between promise and reality from the 1960s, state-led modernisation came under serious scrutiny, along with the assumed nexus between natural resource exports, capital and economic growth.

Oil extraction as development is best captured by approaching it as a network of social relationships, involving a range of individuals and organisations, through which processes of development operate. The form oil development assumes is shaped by ideas and practices acceptable within the networks. Its geographical manifestations reflect the prior social and institutional networks out of which it emerged or to which it is embedded (Bebbington and Kothari 2006: 851).

Oil extraction as trans-local strategic action field

Development occurs within a field of encounters between different actors, national and international institutions, and officials of development agencies, NGOs and discourses (Ribeiro 2002: 169). Suffusing this field are differing political visions, interests and power positions. The array of actors and resources effect a fusion of local, national and international scales. Thus, the process of oil development occurs through networks, which crosses spatial boundaries. Oil development relies on powerful actors like the oil-producing and receiving states, which have a common abiding interest in an uninterrupted flow of the commodity, global financial institutions and multinational oil corporations. In an era of increased Western demand for oil and gas, there is an increased readiness by the United States to deploy military protection of strategic energy sources. The linkages between the scalar levels of interested actors impact on the incidence and character of local conflicts (Dunning and Wirpsa 2004).

Oil resides only in fixed places, necessitating extractive activities at that specific locale. The implication is that oil exploitation generates consequences for the security and livelihoods of communities. Fundamental to the control of oil is the availability of 'infrastructure, security and technology to convert it into asset transportable' (ibid.: 82) across national boundaries. Because oil is simultaneously national and multinational, state oil companies and multinational corporations seek to influence the governance structure, in both the host country and global sphere, which regulate the extraction, production and distribution of oil. The linkages and interactions among local, national and trans-national spaces shape the material interests of competing local actors and the 'discursive strategies upon which they draw to legitimate conflict and militarization' (ibid.: 84). They fault a state-centric focus arguing that the State is just one of many actors attempting to exercise dominion over territories where oil-related violence emerges.

If oil extraction as development churns out benefits for some and costs for others, and has become a contested terrain, it is helpful to describe it as a 'strategic action field' or a social space where two or more organised collective actors engage in conflictual actions (Fligstein and McAdam 1995). Strategic action fields are socially constructed arenas within which differentially endowed groups employing their resources vie for advantage. According to Fligstein and McAdam (1995), the utility of the strategic action field (SAF) lays in its flexibility and the fact that some groups in the action field are themselves strategic action fields. What they perhaps pay little attention to is the view that SAFs could be transnational in scope, in which case they would encompass actors located across spatial scales. This oversight is inherent in the scholars' state-centric focus: 'What distinguishes the State from other SAFs is the distinct claim of its constituent fields to produce or at least ratify the rules for all other fields' (ibid.: 9). Given that oil is an international commodity, its extraction necessarily cuts across spatial scales. Therefore, the trans-local SAF is composed of actors at the local, national and international scales.

Fligstein and McAdam (1995) argue that the first rule in an emergent field or unorganised field is to outline a stable definition of the situation, values and rules guiding relations within the field. Imposition of such rules may come from cooperative relations among the groups or, be imposed by members of a dominant group. Social relations among the field members may be cordial or hostile. Action in the SAF seeks to create and sustain the field in order to ensure uninterrupted flow of group benefits. The rules of engagement that crystallise in the field are 'conception of control' (ibid.), which affirms that the rules are collectively shared cognitive constructs and play the role of controlling interactions in the field. The rules are, however, not benign, nor are they arrived at consensually. To the contrary, they reflect an order imposed by a more powerful or a set of groups that are more powerful. Within the strategic action field, it is feasible to distinguish between 'incumbents and challengers'.

Incumbents are powerful organisations or groups that have the necessary political or material resources to enforce an advantageous view of appropriate field behaviour and definition of field membership on other group. Challengers are organisations or groups that define themselves as members of a given strategic action field, but generally accept the given social order and the advantages it gives incumbents either because they fear retribution by incumbents or because their survivability is increased by accepting such a view. Challengers are those groups who ordinarily exert little control over the field (Fligstein and McAdam 1995: 7).

Conception of control comes into being because of the determined efforts of some groups to fashion consensus on three issues: membership criteria; definition of the goals of the field; and the rules guiding social relations in the field. Efforts at fostering conception of control, the scholars argue, might require the dominant to impose consensus on the less powerful or engineer an encompassing consensus that transcends their own provincial interests. Values and norms in the action field are created through repeated performances such that the order-creating process is always contested and resisted (Henry, Mohan and Yanacopulos 2004). Therefore, there is a need to avoid the materialistic approaches to actors' interests and motivations in organisational studies, which are less attentive to the non-materialistic aspects of networks (ibid.: 2004).

The clash of logics

Western models and approaches have proven inadequate to the task of understanding the environment from the African perspective. Godfrey

Tangwa (2004) contrasts the African environmental worldview to the Western worldview, which is predominantly anthropocentric and individualistic. The author locates pre-colonial traditional African environmental ethics in African metaphysics, or 'eco-biocommunitarianism'; the 'recognition and acceptance of inter-dependence and peaceful co-existence between earth, plants, animals and humans' (Tangwa 2004: 389). The metaphysical outlook explains the benign relations between humans, animals and inanimate objects and supra-natural forces.

Within African metaphysics the dichotomy between 'plants, animals and inanimate things, between the sacred and the profane, matter and spirit, the communal and the individual, is a slim and flexible one' (Tangwa 2004: 389). Against such a worldview, it becomes clearer why African people believe that human beings can turn into animals, plants and, unseen forces, and vice versa. Such a view portends important implications for how traditional African people approach and treat the environment. Ojomo argues that the thin boundary between the holy and profane, human and supra-human and between the communal and individual shaped the traditional African worldview of 'live and let live'. Thus, eco-biocommunitarianism is not metaphysics of domination, consumerism or greed, but ideas and claims rooted in myths and taboos that serve to conserve ecological balance.

The logic of development was different; development planners failed to ask the fundamental question: development for what? Given that 'Development is for whatever men will make of it', Goulet (1968:310) argues that the uncritical acceptance of development goals as desirable for their own sake creates problems.

The problem stems from the fact that the process of engendering development is concerned with how to achieve prescribed targets. It therefore centralises science, and technology. It is not rooted on a normative ethos, which requires choice among goals based on the idea of what the targets of development perceive as right.

Based on a technical order, its emphasis is on appropriate technology to facilitate work or overcome physical barriers in the environment, law and order, mega-finance, and efforts to hunt hydrocarbon and contribute to economic growth. Such technical metaphors depict the environment as 'dead geography' devoid of meanings and values. Worse still, they fail to take into account the interconnection between the environment and its inhabitants. A whole new way of viewing, relating with, and using geography is introduced. The essence of the new forms of territorialisation is to facilitate the attainment of prescribed development goals often defined in quantitative terms.

Leaders of newly independent societies assumed that the goals and processes of development are unambiguously good. Development planners failed to realise that development is a mixed bag of 'goods and bads' (Goulet 1968; Goulet and Wilber 1996). They denied communities cognitive respect by imposing on the latter their own definition of the good society. In that vein, champions of development planning employed a sort of development mantra, 'profitability measure' (Stolper 1966).

According to early development planners, the basis of economic investment decisions should be the criteria of profit and nothing else. They assumed that economic growth would somehow trickle down and percolate every cranny of society. In that regard, little consideration went to the impact of foreign capital on the environment and the effect of environmental change on people and communities.

Goulet observes that development is the pursuit of the good life. He posits what may be considered universal elements of the good life, including 'sustenance of life', recognition, identity, freedom of choice and its expansion. Any process that negates these elements of the good life constitutes pseudo development. And because targets of development experience it either positively or negatively, development is a normative phenomenon.

An understanding of the end of development as economic growth through economic efficiency or criterion of profit leads to the sacrificing of other values that are incompatible with profit and efficiency. Thus, cultural values have been destroyed on the altar of development. Communities continue to pay a gratuitously high price for development. How development is achieved has meant little to its paragons.

Very perceptibly, Goulet avers that choices as to whether to pursue development, through what means, at what rate, and what level of human costs was acceptable were 'tantamount to votes cast in favor of a particular social system, a philosophy of life, a universe of meaning' (p. 304). In this case, the vote was cast in favour of a technical order, a system that churns out benefits for global capital based on deliberate coercion or exploitation of local communities through the means of science geared at economic growth.

Thus, oil development in Nigeria described in the metaphors of 'oil extraction', and 'oil extraction as translocal strategic action field' rises on a technical order. By technical order I refer to that 'which results from mutual usefulness, from deliberate coercion, or from the mere utilization of the same means' (Redfield 1953: 20–21).

> This is why definitions of development in terms of industrialization, urbanization, modernization, growth, maximization, or even optimization are bad definitions. If we wish to speak of development comprehensively, we must speak of it normatively.
>
> (Goulet, 1968: 309)

Amoral development or development not based on any moral order but a technical order, is driven and sustained by exploitation of communities and land for the elites' benefit. Amoral development come into conflict with local community's moral order in which people value nature as living and part of human community.

By means of State power disguised in the form of laws for the national good, discourse of development and clientelistic ties with local elites in the region, the technical order predominated over the moral order. The territorialising logic of the technical order effectively nipped the territorialising capacity of local communities. Worse still, outcomes of the territorialising patterns of the technical order served not only to penalise local communities but to disrupt and undermine bio-communitarian ethos. The incommensurability between the technical order as represented by capital and the moral order represented by pre-existing mode of being and relating with the environment began to provoke counter-discourses and reaffirmation of the moral order.

Conflict in the trans-local action field

This section presents a conceptualisation of development/oil extraction as fields of trans-local strategic action. A system or functionalist perspective would suggest that such a field is well integrated, its various parts functioning harmoniously to produce the desired goal of resource extraction and development while keeping all parties happy ever after. However, the field of strategic action is composed of processes of integration and disintegration, stability and conflicts, benefits and costs. The news of the commencement of extractive activities may generate opposition or excitement among the would-be stakeholders based on expectations. Beyond these potential initial responses, the extractive industry, as a network of relations, induces conflicting experiences, interests and visions of social organisation especially with regard to resource production, resource allocation, distribution of benefits and costs, environmental risks, environmental management, resource control, the nature and costs of development and the relationships between firm and stakeholders (Albrecht, Amey and Amir 1996).

The development process generates contradictions and 'polarization between functional elites and the functionally superfluous' (Apter 1993: 3). Where development induced displacement and conflicts resonate with problems of socioeconomic marginalisation and poverty in the wider society, the hardening of differing positions and intensity of conflict assume dreadful dimensions. According to Apter, the functional elites organise capital-intensive production methods that engender the marginalisation of those who become functionally superfluous. Such production techniques contribute to the large-scale transformation of the physical topography, which in turn imperil the livelihood of the land dependent community. Priority goes to sustaining uninterrupted exploitation and supply or conditions favourable to capital accumulation over unemployment, local livelihood, social, cultural and environmental effects of development. As Apter emphasises, the political system is least responsive to the marginalised, occasioning the 'invisibility' of the latter. It is within such contexts that emancipatory projects begin to emerge.

Functional elites may attempt to protect their 'privileged access' and 'privileged accounts' by arguing the benefits their presence or operations provide the field and entire economy, and how any adverse form of intervention in the status quo might affect the economy (Freudenburg 2005). Moreover, the elites might resort to 'diversionary reframing' as a strategy of changing the terms of the debate (ibid.: 104) in which strenuous effort is made to dent the credibility of challengers or directly point at something else other than what challengers named as the object of their grievances. Furthermore, elites maintain their privileges through the social construction of 'quiescence or *"non*-problematicity"' (ibid.: 105). Situations and events described by challengers as displacing and destructive are energetically constructed by the elites as non-problematic, amenable to resolution, and/or defined as emanating from something other than the operations of the elites.

Collective actors in the trans-local strategic action field

To Apter (1993), people penalised by development interpret their negative conditions with a view to transcending those circumstances by thinking beyond them. They achieve this through mytho-logics: turning events and experiences into stories and myths, which are in turn explained by means of logical principles. The collective action that develops is as much for as against. Thus,

> Theirs is the politics of the moral moment, disjunctive, redemptive or transformational. Claiming legitimacy against current principles as well as excesses of power, the defects of society are interpreted as failures of the state. Movements like these arouse controversy by their very existence and stimulate debates over political fundamentals. Their chief weapon is a discourse capable of threatening prevailing norms and principles of power particularly when combined with confrontational episodes.
>
> (Apter 1993: 12)

Such movements are least concerned with rectification of inequalities and exclusions as in undermining codes and discourses. That attribute, Apter argues, is what separates them from the 'old' social movements, which allegedly fought for greater participation and equality.

The New Social Movement (NSM) approach gives weight to the overriding importance of structural conditions in the emergence of social movements. The approach questions the economic and class reductionism of classical Marxism, arguing that non-class issues such as the environment, gender and peace, rather than economic changes and the class position of actors, explain the emergence of social movements. For instance, Habermas (1987) distinguishes between the *life-world*, and the State and market. While communicative rationality gives the life-world structure, instrumental rationality is the structure of the State and market. The expanding processes of instrumental rationality inundate the life-world, continually absorbing it; a process he termed the colonisation of the life-world. Social movements, Habermas argues, are the result of such colonisation; they arise in reaction to colonisation of the life-world and seek to recreate lifestyle. Laclau and Mouffe (1985) attribute social movements to changes in the social structure. The Fordist mode of production engendered fundamental changes in production, in the nature of the State and culture itself, which resulted in increased commodification, bureaucratisation and massification. The increased penetration of wider spheres of social life by capitalist relations has led to the transformation of society into a big marketplace, or commodification. Social movements arise to challenge such processes.

Conflict in Translocal Strategic Action Field (TSAF) sits in place

Tilly (2000) argues that space relates to contentious action and vice versa in five ways. 1) Conflict happens in places occupied by people, in which case the spatial configuration may facilitate or hinder participation in collective action. 2) Everyday spatial routine, spatial distribution and proximity shape patterns of mobilisation. 3) Territoriality organises space for government, disrupted by collective challenges. 4) Routine political life, including protests and public ceremonies endows some places with symbolic significance. 5) Contention transforms the political significance of given places and spatial routine.

A major failing of approaches to conflict and social movement within the context of development is the near total disregard of space. In their conceptualisations, one perceives the unmistakeable impression that space is a mere background against which social events occur. Thus, in Touraine's (1985) definition of conflict, he disregards how place may be a component of the antagonist actors' identities and that a sense of place may thoroughly structure the prized object of contention. The role of place in generating emotions and mobilising collective action remains largely neglected in the resource mobilisation school. Place is more than geographical context and dead geography; it is meaningful and symbolic, and is thus entwined with the social activities of those who live in it. However, place is equally the outcome of processes operating at wider scales. Therefore, sensitivity to the wider space composed of distant and local actors whose activities affect the local place is vital. As presented here, one may view development as a wider space. Yet, that view omits important considerations.

Colonial development engendered unequal regional development or spatial inequalities. Postcolonial national development has widened regional inequalities, making some places more politically strategic than others are (Okafor 1980). Given that elites of the dominant groups formed the core of the ruling elites, they use their political influence to initiate and locate development projects in their regions more than in minority places. Exercising political power, the elites employed constitutional means to organise, appropriate and distribute oil resources in favour of dominant interests and to urbanised places in dominant regions at the expense of minorities, including oil-producing communities. Worse still, oil development combined with expropriation to degrade and pollute the environment of oil-producing communities with the effect that the disparities between dominant and peripheral places have widened. Beyond environmental degradation, the connection between places and those who inhabit them degrades. Thereby, the meanings, values, inspiration and sense of attachment or ghosts people invest their place with are denigrated and profaned.

It is impossible to explain why and how conflict emerges in the trans-local strategic action field, without addressing how development furthers and sustains spatial inequalities, enhances the opportunities of some and not others because of where they live and how these processes affect the bond between humans and non-humans and the sense of place people hold dear. Oil development practices seem to sustain relations of inequalities between peoples and places, and given the bond between environment and community, conflict within the strategic action field is about spatial inequalities as it is about social inequalities.

Mittelman (1998: 848) emphasises the need to defy ontological division between human and non-human entities. Cultural theory draws attention to the interdependent relationship between human communities and the other-than-human world in which they are situated. Conceptualising the nexus as 'multiple articulations of community', Whitt and Slack (1994: 21–22) argue 'community and environment constitute a single integral and open system'. The authors eschew anthropocentrism, and argue for bringing the human and non-human together in 'relations of solidarity and significance'. Such relations evoke a view of community as a 'unity in difference' rather than as a 'unity of sameness', in contrast to a conservative understanding of community (Young 1990). They say the concept of community makes sense because the processes of subject formation take place in communities. Moreover, community mediates the salience of global forces, ostensibly because it is within community that hegemonic oppressive forces are experienced and resisted (Whitt and Slack 1994: 8). Failure to extend communal relations of significance and solidarity to other than humans 'is central in most environmental problems' and has worsened social conflicts (ibid.: 19).

Conclusion

This chapter argues that since, in conventional development, resource exports traditionally provide the fastest means to growth for poor countries, development should be conceived as oil extraction. Proceeding from the premise that development is inherently conflictual, it observes that the best approach to such development is not as impersonal phenomenon or structure but as a process involving identifiable actors and associations among people and places across spatial scales. These actors have interests, which are sometimes complementary and most often conflictual, which they attempt to realise by manoeuvring other actors. In other words, the externalities of development and the costs they impose on the less powerful are not accidental or fleeting. To the contrary, they adhere to development itself. However, the conflicting actors in the field do not separate into homogenous entities. Rather, collaboration among elements of conflicting groups of actors does exist. To capture such dynamics, the metaphor of development as trans-local strategic action field is utilised. The conflict in the field is conceptualised as social conflict, defined by three elements: identity of the protagonist, the opponent and the stake over which both struggle.

Drawing on Touraine (1985), this chapter underlines the problem of viewing conflict in SAF as a mere reaction to structural conditions. In other words, one should not analyse conflict utterly as a facet of a social system, but in relationship to conflicting actors and a stake. Such conflicts may involve mobilisation of people, identity construction and the assemblage of resources. It is best to capture collective actors in SAF by asking why they mobilise and how. To explore that question, this chapter adopts social movement theories. It goes further to argue for emplacement of that social movement because place mediates collective action. A social movement, its impact, nature and trajectory are shaped by the environment in which it emerges. Nevertheless, social movement mobilisation involves mobilisation of actor-spaces or the engineering of association of actors across spatial scales in the effort to realise movement goals in a given location.

5   Why the Ogoni mobilised

Introduction

On 4 January 1993 about 300,000 Ogoni converged, under the auspices of the Movement for the Survival of Ogoni People (MOSOP), to take part in an unprecedented protest march against the State and Shell Oil Company (Saro-Wiwa 1995). The Ogoni located their grievances in a skewed federal structure, environmental despoliation and destruction of local livelihoods by the activities of Shell (ibid.). Threatened and angered by the MOSOP, the State – actively aided by Shell – harassed and intimidated Ogoni leaders (Ibeanu 2000). In November 1995, the State executed Saro-Wiwa and eight other Ogoni leaders on trumped up charges of masterminding the murder of four prominent pro-government Ogoni chiefs. The judicial murder sparked a wave of international outrage and condemnation (Maier 2000). The sudden death of Nigerian leader Sanni Abacha, and subsequent election of Olusegun Obasanjo in February 1999 ended Nigeria's international isolation and held the promise of a quick and peaceful resolution of the Niger Delta imbroglio.

Contrary to expectations of peace, several militant groups have emerged (Ibeanu 2000). The violent attacks and activities of the militant Movement for the Emancipation of the Niger Delta (MEND) resulted in roughly a 25 per cent cut in Nigeria's crude oil production in 2007. Hostage-taking of foreign oil workers, seizure and destruction of oil installations, and armed confrontations between federal troops and militant groups have become common. It is alleged that the policy sources of grievances and conflicts – such as the Land-Use Act, Petroleum Act and other laws that dispossess and marginalise oil-producing areas – coupled with the operating practices of oil companies have not yet been adequately addressed by the State nor by the oil companies (Douglas, personal interview 2006).

The conflicts in the region have engaged sustained scholarly attention. Several approaches to the Niger Delta conflicts can be discerned in the literature. While the authors present important insights into the conflicts, little regard has been given to place-specific factors that determine ideational reality, the value of resources, and the power relations that organise such resources (Peluso and Watts 2001).

Conflict over a region's natural resources is a geographic phenomenon (Simmons 2005). Political geographers have long recognised the role of space and place in the emergence of conflict (Miller 1994; Agnew 1987; Pile and Keith 1997). Some social movement scholars have embraced the spatial perspective (Sewell 2001; Martin and Miller 2003; McAdam et al. 2001). Space is essential to the choice of tactics and strategy (Agbonifo 2009) and construction of frames and identity (Agbonifo 2009; Wolford 2003). The insight has, however, had minimal impact on scholarly engagement with the Niger Delta conflicts. This chapter moves beyond existing literature on the conflict in two ways. First, it seeks to situate structural conditions in place. Second, the chapter traces how interpretive activities, partly informed by place, transformed conditions to construct insurgent identities. Building on insights from 'resource access literature', 'society-rooted politics' and place-sensitive social movement theories, it argues that violent conflict emerges at the intersection between structural conditions, place-specific characteristics and place-informed interpretive activities.

But first, there is a need to outline what we mean by social conflict to enable us trace how the dominant macro-explanations of the Ogoni conflict engage with it.

What is social conflict?

Conflict may be seen as the outcome of longstanding structural inequities, environmental degradation, contradictory securities, and attempts to secure a larger share of the national pie. But such definitions breach Touraine's canon that a social conflict should not be analysed utterly as a facet of a social system. Conflicts are marked by a clear definition of protagonists and antagonists, and the resources or stakes they fight over. In other words, actors in conflict are organised and oriented to goals considered valuable by the conflicting actors. Thus, he observes that any conflict has three elements, namely the identity (i) of the actor, the definition of the opponent (o), and the stakes (t), which defines the field of conflict.

Touraine identifies three categories of social conflict: firstly, the competitive pursuit of collective interests, and the reconstruction of a social, cultural or political identity; secondly, a political force seeking to change the rules of the game. The competitive pursuit of collective interests within an organisation may degenerate to conflict over organisational rewards. This is likely when high output is lowly rewarded. The reconstruction of identity responds to an organisational status and organisational change. The actors seek to defend their positions from perceived external threats. Stakes for the political force are the rules of the game, and not merely the distribution of advantages in an organisation. Thus, the actors define themselves, the opponent, and the rules that need to be changed or retained. Lastly, the conflict whose stake is '*the social control of the main cultural patterns*, that is, of the patterns through which our relationships with the environment are normatively organized'

(Touraine 1985: 754–755, emphasis in original). It refers to the conflict between hegemonic deployment of knowledge, investment and ethical principles and the redefinition by the masses of representations of truth, production and morality.

Each of the three levels of social conflict, Touraine argues, can be distinguished by the degree of integration among the three elements of social conflict; definition of the identity of the actor (i), the definition of the opponent (o), and the stakes, which define the field of conflict (t). At the level of social control of main cultural patterns, the components are integrated, homogenous and interdependent (i-o-t). The political struggle to reconstruct identities is a different type of conflict in which the three elements are less integrated because there is no direct interdependence between political forces and decision making systems. In effect, the elements of the social conflict, or political struggles, can be written as i-t, o-t, or i-o. The competitive pursuit of interests in organisations manifests loosely integrated elements. The actors are self-centred and the field of their conflict can be defined as a market, or independently from the actors. Non-integrated and separated from one another, the elements of the social conflict stand on their own as i, o, t.

As Touraine (1985) points out, the subsystems collective action is directed at tells us about the nature of the conflict. In the case of the Ogoni, contentious activities were directed at subsystems responsible for production, appropriation, and redistribution, the rules governing the foregoing, and ideas of the good society. Thus, we have simultaneously several social conflicts, some materialist in orientation and others cultural. As we will see later, some conflict analysts are not able to disentangle the different strands of conflict with the result that a complex event is often reduced to the noticeable aspect of its elements.

A study of collective action must differentiate between a reaction to strain or crisis and the expression of conflict (Touraine 1985). When seen as an effect of systemic crisis, social movement is reduced to systemic pathology. Social conflict is, however, defined by a struggle between two or more actors striving to capture or control resources considered by each of them as valuable. Therefore, for an event to constitute a conflict, 'the actors must be definable in terms of a common reference system, and there must be something at stake' to which they are oriented (Melucci 1996: 22). By failing to distinguish between crisis and conflict, we are unable to understand forms of collective action.

Thus, Melucci correctly observes that a collective actor is a complex phenomenon and operates within an equally complex terrain,

> A collective actor operates within various organizational systems at once; it lies within one or more political systems; it acts within a society comprising various coexisting modes of production. Its action therefore involves a whole range of problems, actors, and objectives.
>
> (Melucci 1996: 37)

However, one dimension of the conflict may predominate and give the movement its essential character.

Moreover, conflict analysis needs to distinguish among different orientations of collective action. Some collective phenomena are the aggregation of atomised behaviour, which involves no solidarity. It forms in reaction to crisis or rapid change. Some collective phenomena come into being through conflict, while others are through consensus. However, there are collective phenomena that breach 'the limits of compatibility of the system of social relationships within which the action takes place' (Melucci 1996: 24). When a conflict respects the limits of its reference system, then action merely seeks reform within the system. Analysis that introduces the notion of transgressing boundaries must define a reference system. A system can be seen as a complex of relationships among its constituent elements. Melucci, thus, characterises systems according to the types of relations composing them. These are: first, system that ensures production (consists of antagonistic relations over production and appropriation and distribution of societal resources); second, system that decides on resource distribution (refers to the political systems, which make normative decisions); third, system of rules governing exchange (relationships aiming at system equilibrium, and adaptation through integration and exchanges); and fourth, lifeworld or system of social reproduction (refers to where basic requirements of social life are reproduced and sustained).

In consequence, Melucci defines a movement as follows:

> A movement is the mobilization of a collective actor (i) defined by specific solidarity, (ii) engaged in a conflict with an adversary for the appropriation and control of resources valued by both of them, (iii) and whose action entails a breach of the limits of compatibility of the system within which the action itself takes place.
>
> (Melucci 1996: 29–30)

Thus, a movement may be distinguished according to its field of action: conflictual networks; claimant movement; political movement; and antagonistic movement. Conflictual networks refer to conflict and breaking of rules at the lifeworld level. Here, action is directed at the rules governing social reproduction in everyday life. Claimant movements press for 'a different distribution of resources within the organization', thereby clashing with the power behind subsisting rules of distribution (Melucci 1996: 35). Antagonistic movement is collective action aimed at production of society's resources; it questions how such resources are produced, the objective of social production and the nature of development. A political movement seeks to 'improve the actor's influence over the decision-making processes, or to ensure its access to them' (Melucci 1996: 35). To Melucci, these distinctions are important because the dominant tend to

> deny existence of conflicts which involve the production and appropriation of social resources. At the very most they acknowledge the existence

of grievances or political claims, seeking however then to reduce all conflictual phenomena to these only.

(Melucci 1996: 35)

The point should, however, be made that the various referent systems are not so distinct and autonomous in reality. The boundaries among them are fuzzy and they are clearly interdependent. Thus, the distinction serves mainly analytical purposes, helping us tease out insights that might otherwise remain hidden. Similarly, the value of the different conflict orientations lies in its analytical insights. A given conflict may comprise two or more of the orientations, and given the fuzziness among the boundaries of the various orientations, maintaining such distinctions in reality may be misleading.

Dominant macro-explanations of Niger Delta conflicts

This section considers several approaches to the Niger Delta conflicts. The approaches are briefly treated under two broad and arbitrary categories: (1) resource availability and (2) society-rooted politics. It is important to point out at the outset that both approaches are convincing at the macro level, but they do not explain why individual actors joined the Ogoni movement because they do not actually situate the conflict in place or people. As Wolford (2003) argues in the case of the Movement of Rural Landless Workers in Brazil, conventional macro explanations mistakenly assume a direct link between broad structural changes and mobilisation. Unable to explain who joined the movement, or did not, and why, conventional approaches remain 'thin on the internal politics of dominated groups, thin on the cultural richness of those groups, thin on the subjectivity – the intentions, desires, fears, projects – of the actors engaged in these drama' (Ortner 1995: 190).

Resource availability

There are two schools of thought on how resource availability shapes conflict: first, scarcity results in conflict, and second, resource abundance shapes conflict. The scarcity school argues that scarcity occasions conflict as groups contend for access to limited resources. Conflicting environmental security perspectives, environmental degradation, overexploitation of resources, population growth and climate change and population movement are some of the processes posited to engender resource scarcity and precipitate conflict (Ibeanu 1997; Homer-Dixon 1999). The school has been criticised by those who argue that it is precisely resource abundance (alternatively, resource curse) that shapes conflict (Le Billon 2001; Watts 2001).The resource curse literature thus draws a link between resource abundance and conflict. Collier and Hoeffler (2002) argue that there is a correlation between poverty, natural resource abundance and violent conflict. Collier (2001) is convinced that countries dependent on resource exploitation seem to be among the most conflict-ridden countries in the world. Greed has also been emphasised (Collier 2001; Reno 2002; Watts 2004; Omeje 2006). What circumstances create the context for resource abundance and competition? Failure to explain those contexts of resource abundance and competition by resource curse literature elicits Watts' critique that an approach based on resource determinism de-emphasises politics as an essential explanatory variable (Watts 2004: 53). Taken together, the resource availability or abundance provides insight into the Niger Delta conflicts as follows.

The approach reiterates that land is very scarce in the Niger Delta and the little that is useable is treasured, forming the spiritual and material basis of life in the region (Obi 1999). In this region, and particularly in Ogoniland, both the land and rivers are central to all economic, social and domestic activities. Given that about 90 per cent of the total Niger Delta area consists of water, from many centuries in the past, canoes were critical and indispensable to movement, communication and trade. Long-distance commerce required large canoes to convey bulky goods (Kpone-Tonwe 1997: 25). As timber for construction was rare in the Delta (which is a mangrove terrain), canoe-building centres sprung up in areas such as Ko village in Ogoni, which had thick forests and timber. There was a concentration of huge wealth in Ogoniland due to expanding canoe and pot industries and bountiful farm harvests. The accumulation of goods created the problem of storage, and according to Kpone-Tonwe, as a result of this, the Ogoni converted their wealth into other forms of wealth: land, permanent tree crops such as palm oil and coconut trees and money (ibid.: 131). By the sixteenth century, a class of wealthy men, whose wealth derived from commercial enterprise, had emerged in Ogoniland (ibid.: 34–36).

Until the end of the nineteenth century, the plain of Ogoniland was densely forested. The fertile plain ensured that Ogoniland became the food basket of the Niger Delta. The Ogoni produced provisions taken on board slave ships (Kpone-Tonwe and Salmons 2002). With population growth and increased demand for farm produce from the Delta, the early twentieth century witnessed the conversion of large areas of forest into farmland. Accelerated population growth increased pressure for farmland such that even the wetter areas of the land were cultivated for quick cassava crops in the dry season, endangering valuable water resources and impoverishing the soils (ibid.: 275). With a population of 500,000 people squeezed into 404 square miles, and an estimated population density of 1,250 persons per square mile, the question of land is a very sensitive issue to the Ogoni (Obi 1999). Thus, as Obi argues, land scarcity and environmental degradation are at the core of the struggle in the Niger Delta (ibid.).

A new current that would exacerbate land scarcity and degradation was introduced with the first oil discovery in 1958 in the Ogoni community, Kegbara Dere. Shell Petroleum Development Company (SPDC), in a joint venture partnership with the Nigerian National Petroleum Corporation (NNPC), Elf and Agip operates five major oil fields and 96 wells, linked to five flow stations in Bomu, Bodo West, Ebubu, Korokoro and Yorla, all Ogoni communities (Banjo 1998). In 1914, the colonial state passed the Mineral Acts declaring sovereignty over mineral resources, empowering the Governor-General to grant licences and leases to British companies and subjects. In 1938, Shell obtained rights to prospect for oil in the entire Nigerian land space. The company later concentrated on an area of high expectation measuring 15,000 square miles and returned the remaining land space to the colonial state. Shell drilled its first oil wells in 1956 and began oil export in 1958.

It took the postcolonial State nine years after independence to repeal the 1914 Mineral Act and enact new oil-related legislation. Shell continued to operate freely within the favourable institutional framework crafted by the colonial state almost a decade after independence (Frynas 2000). Even then, the 1969 Ordinance was little different from its colonial antecedent (Frynas 2000: 81). Between 1971 and 1990, there was no formal operating agreement between the State and Shell. For two decades, Shell operated within an institutional void and without obligations (ibid.: 89) ostensibly because the Mineral Acts placed responsibility for oil exploitation in the hands of two monopolies: British Petroleum and Royal Dutch Shell. The monopoly rested on their agreement with the colonial State to share oil proceeds 50–50 (Osoba 1987).

The Ogoni celebrated the discovery of oil with excitement and hope (Agbo 2008). The installation of oil facilities generated jobs for unskilled labour and attracted small service sector industrialists, job seekers and other migrants to the region. The processes of urbanisation increased, Ogoniland boomed, and the people were excited. However, as Gaventa (1982: 56) argues in the case of the Appalachian Valley, below the surface of the boom, the legitimacy investors enjoyed and the 'momentary Zeitgeist, there was quietly occurring the structuring of inequalities that was to have major long-term impact upon the political economy of the region'. The 'oil rush' of the early 1970s unleashed rapid land alienation, quickly resulting in mass landlessness. The massive dispossession or 'material haemorrhage' that became a characteristic feature of the Niger Delta came about through forcible expropriation, deceit, corruption and state acquisition (Otite 1990: 327, 332). By the early 1970s, the hope in oil was dashed and in its place, a grim realisation of despoliation settled (Agbonifo 2003). Ogoni leaders resorted to petitioning the State.

The ecology of Ogoniland has undergone profound changes as a result of oil-related activities (Boele et al. 2001). The estimate of carbon dioxide (CO^2) emissions from gas flaring in the Niger Delta stands at about 35 million tonnes annually, the highest annual emission from gas flaring in the world. The large volumes of greenhouse gases released, such as CO^2 and methane, contribute to global warming (Orubu et al. 2004). The soot released causes acid rain, fouling bodies of water and destroying once fertile farmland. Constant gas flares negatively affect the environment, destroying plant growth and wildlife, and driving away important species. Farmers harvest less returns yearly despite hard work and the Ogoni now must buy food from outside (Amanyie 2001: 18).

Society-rooted politics

As if to underscore the fears and vulnerability of the Ogoni, in July 1970, a blowout occurred in the Ogoni town, Dere, where Shell first struck oil in 1958 (Saro-Wiwa 1992). According to one report from Dere, 'The blow-out continued day and night for about two months during which we were forbidden to make fire, we could neither cook our meals nor smoke tobacco' (Saro-Wiwa 1992: 72). So severe was the disaster that it destroyed farmlands within a radius of about three miles. Worse still, the blowout occurred during the harvest period, destroying the first fruits after the civil war. Yet, not a single relief material was received in Dere, as the victims 'were left to swim or sink within their miseries' (Osha 2006: 28). Saro-Wiwa charges that what Shell has done to the Ogoni people, land, streams and atmosphere amounts to genocide, murdering the soul of the Ogoni (Saro-Wiwa 1992: 75).

In a petition addressed to the governor of Rivers state on 25 April 1970, Ogoni leaders alleged that after Ogoni returnees, displaced by the civil war, had been encouraged to till the land to eke out subsistence, Shell–BP trucks and earth-moving vehicles entered cultivated farmlands and bulldozed several acres of crops (Saro-Wiwa 1992). Prior to the petition, the leaders had shown the governor acres of mangrove swamps destroyed by incessant oil spills, imperilling the livelihood of the poor. Crude oil and mud polluted the once sparkling rivers and streams in Gokana area, leaving the people no alternative source for drinking water. Saro-Wiwa (1992: 47) summed up the Ogoni travails thus: 'Our people have been compelled to sacrifice all life-supporting necessities so that the nation may enjoy economic boom'.

Shell–BP countered the Ogoni accusation, arguing that the charges are inaccurate (ibid.: 50–51). Nineteen years after the Bomu blowout, some families who were victims of the blowout sued Shell for compensation in *Shell v. Farah*. In March 1988, Shell claimed it had rehabilitated and handed the land over to the plaintiffs and that it had paid compensation for damaged trees, crops and other elements as well as for land degradation (Frynas 2000). The plaintiffs claimed ignorance of Shell claims and consequently initiated legal action in 1989. In court, Shell relied on an expert witness (Frynas 2000: 168). The judge awarded the plaintiffs Naira 4,621,307 (US$29,924.9) in compensation. Shell appealed the judgment but it failed before the Court of Appeal. However, such expert claims have in some cases helped Shell avoid responsibility for environmental damage (ibid.: 169).

Perhaps as a result of its dominance, Shell was unperturbed by the increasing number of conflicts, which dogged its operations. Frynas (2000) shows that in the period 1981–1986, Shell had 24 compensation claims lodged against it in Nigerian courts. In 1998, the number jumped to more than 500 cases. Most of them were oil-spill related. Shell claims it is a victim of sabotage and has employed the guise of sabotage to escape liability for damage. However, Shell has so far failed to take legal action against any suspected saboteurs.

With land degradation and alienation, it was only a matter of time before an underclass emerged whose impoverishment could be tied to the lack of adequate regulation of multinational oil companies and subsequent destruction of the local economy (Agbonifo 2003). In 1990, the Ogoni took stock of their condition and concluded that they were 'faced with environmental degradation, political marginalisation, economic strangulation, slavery and possible extinction' (MOSOP 2004: 2). When, as a result, the Ogoni issued its Bill of Rights and ultimatum to Shell demanding payment of compensation, both the State and Shell ignored the movement (Ibeanu 2000). Following the rise in contentious activities, the State issued a decree, which criminalised any call for autonomy (ibid.). On its part, Shell commenced a campaign against the person of Saro-Wiwa, arguing he did not represent the entire Ogoni (Saro-Wiwa 1995). Moreover, the company attempted to impugn the environmental claims of the Ogoni by arguing non-problematicity (ibid.).

The Ogoni quest for social justice and equality reflects the broader problem encapsulated in the National Question. The core of the National Question relates to how people are organised, empowered or disempowered (Momoh 2002: 26). Osadolor (2002: 31) argues the National Question arose from the amalgamation of the Southern and Northern Protectorates in 1914, the subsequent incapacity to transform the complex into national societies and the consequent problem of what to do with the country. Colonialism engendered divisive policies and made little effort to create a united country. Some colonial officials did not believe Nigeria constituted a single country and expressed a lack of faith in the entity they had created (ibid: 32). These forces fostered and enforced the feeling or perception of difference, fear and suspicion.

Mistrust persisted even after independence in 1960. Politics degenerated into a struggle for power at the federal level. Possession of the reins of power at the centre assured access to economic survival and benefits as well as other social ends. In their confrontation over sectional goals, rival groups dispensed with self-restraint leading to a series of political crises that resulted in Nigeria's first military coup on 15 January 1966. The ensuing civil war, the outcome of which favoured federalism, did not resolve the National Question but merely imposed unification (ibid: 45). The primary source of crisis in the post-war era has been the inequitable distribution of national resources, in which ethnic minorities of the oil-rich Niger Delta question the essence of Nigeria and advocate convening a sovereign national conference to debate continued coexistence (ibid.: 43–4).

Extant inequalities in the distribution of wealth generate instability and protract the National Question (Anikpo 2002: 66) which, in effect, is about the issue of equity with regard to resource distribution among the various ethnic and class groups that compose Nigeria. Interethnic inequalities and the National Question predate the emergence of oil as a major revenue earner for the country (Obi 2002: 97). The politicisation of interethnic relations, expressed by the National Question, led to the struggle of majority groups to maintain domination at the expense of the minorities. The tendency of the latter was to escape their domination by opting out of a 'contract of perpetuity in inequality', an option the dominant group actively resisted (ibid.: 98). Minority fear and protests against majority domination led to establishing the Willink Commission, which failed to address minorities' anxieties (ibid.: 99). As a result of minority marginalisation, local elite anger and rural dispossession 'exploded' into Ogoni militant nationalism in the 1990s (Okonta 2008: 5).

The resource availability approach provides a materialist explanation of the Ogoni conflict, which involves elements of grievances and greed theories. The political economy of conflict argues that both grievances over exploitation and greed to acquire more resources for provincial needs explain why people resort to contentious politics (Gurr 1970; Collier 2001). The society-rooted approach privileges a materialist and provincial explanation of the Ogoni movement, which aligns with the grievances theories founded on the argument that arduous economic conditions, or perceived inequality, compel resistance to unfavourable conditions. Both the resource scarcity and society-rooted politics perspectives emphasise the critical role played by the movement organisation, Movement for the Survival of Ogoni People (MOSOP) and leadership in the rise of the mobilisation. Assent on institution and leadership reflects the resource mobilisation standpoint, which underlines that the capacity to assemble resources determines whether, and where, a movement forms.

Political opportunity

Elite participation in the struggle, which hints at elites' division at the national level, was a political opportunity. But it was also a political liability. It was a political opportunity because it gave the movement the kind of seriousness that it would have lacked had the elites decided to ignore the movement. It was a liability because the elites sought to utilise the resource of the movement in an accommodationist way. While the movement created space within which to rattle the State, the elites sought to steer such advantage in a way that would not endanger its clientelist relationship with the State. The vast majority of the Ogoni aspired to utilise the traction created by the movement to alter or abolish relationships of domination.

We need to address what is political opportunity. And to whom it is an opportunity. Political opportunity should be related to movement goals. Is it an opportunity to resolve some tangential goals while leaving the main goal unmet? Is it an opportunity geared towards cosmetics or fundamental change? What happens when there is a struggle to direct the opportunity in two conflicting directions? What if the apparently existing opportunity merely supports reforms and not fundamental change as some desire? Within the Ogoni movement, three types of conflicts were envisaged: conflict involving competitive pursuit of the marketplace, conflict to alter the rules of the game, and conflict to control the main cultural patterns. While the elites were interested in the first two, ordinary Ogoni prioritised the latter. In their reckoning, the first two conflicts are embedded in the latter; resolve the larger issue and the minor ones would be resolved.

Blindspots of dominant explanations

Table 5.1 below shows several things. First, it indicates that analysis that restricts itself to considering the MOSOP only in relation to distributional matters, or political matters, or anyone of the four systems Melucci itemises is inadequate. Thus, the tendency to conceive the conflict as environmental, ethnic or self-determination merely essentialises the movement. Second, the mix of Ogoni demands and the systems of reference show the intermingling of factors normally checked into binary opposites: recognition and redistribution. Such intermingling does not call for an understanding that asserts that the movement had dual motivations. What it suggests, I will argue, is that each demand has both components, the same way that each system is an entanglement of recognition and distribution. Importantly, the table helps us to see that the Ogoni conflict is more complex than has been treated in the literature. It engages with surplus production as with social production, it concerns redistribution as with participation in the rules and decisions governing redistribution.

Recognition and redistribution?

A pessimistic slur has been thrown on the MOSOP and its leader, Saro-Wiwa, by some scholars and activists who argue that self-oriented interests or provincialism explain the Ogoni motivation. Many authors claim that Ogoni demands are traceable to the greedy appetite of Ogoni elites for a greater share of the national cake (Reno 2002; Watts 2004).

Reno (2002) argues that the Ogoni aimed at inclusion in the status quo. Lacking practical economic prospects, the Ogoni mobilised for cooptation into existing patronage networks as a means to improving their condition; they did not seek to undo the system that rendered them superfluous and invisible (Reno 2000: 50–51). Okome (2000) accuses Saro-Wiwa of narrow and virulent Ogoni ethnocentrism by failing to incorporate all other marginalised Niger Delta communities (see Pegg's (2000: 701–708) rebuttal). A legacy of MOSOP is the bitter and violent interethnic struggles over territory, and such is the case because MOSOP made 'the politics of territory and property of central concern' (Watts 2004: 71). Watts' argues that MOSOP generated a 'space of indigeneity', resulting in a recapitulation of the 'post-colonial history of spoils politics in Nigeria' (ibid.: 291). Leton described Saro-Wiwa as one who was willing to sacrifice everything to secure his selfish desires (UNPO 1995: 15).

These representations and insinuation of political motive have been constructed despite Ogoni activists', including Saro-Wiwa, self-understanding of their motivations. For instance, Saro-Wiwa claims that his quest is for social justice, not the break-up of Nigeria. To enable us to resolve the contradiction between activists' self-understanding and theoretical inputations of motives, certain pertinent questions need to be addressed.

Where did Ogoni demands on the Nigerian State as enshrined in the Ogoni Bill of Rights emanate from? The Ogoni made seven clear demands on the State as shown in Table 5.1 above. Reflection on the question of the roots of the demands is important in light of the misleading argument or suggestions in the literature about Ogoni claims making. As with the mobilisation of

Table 5.1   Varieties of conflicts Ogoni claims represented

Ogoni claims/demands derived from the OBR	*Systems claims are directed at*	*Systems/claims mix*
1. Right to control and use of fair amount of Ogoni economic resources	a. Systems of production b. System of resource distribution c. Systems of rules governing exchange (redistribution and recognition)	a. 1 b. 1, 2, 3, 4, 5, 6 c. 1, 2, 3, 5, 6, 7 d. 4, 5, 6, 7
2. Political control of Ogoni affairs by the Ogoni	b. System of resource distribution c. Systems of rules governing exchange (redistribution and recognition)	
3. Representation in all national institutions	b. System of resource distribution (redistribution and recognition) c. Systems of rules governing exchange	
4. Protection of environment	b. System of resource distribution d. lifeworld or system of everyday life reproduction (redistribution and recognition)	
5. Development of Ogoni languages	b. System of resource distribution c. Systems of rules governing exchange (redistribution and recognition) d. lifeworld or system of everyday life reproduction	d. 5, 6, 7 b. 5, 6 c. 5, 6, 7
6. Development of Ogoni culture	b. System of resource distribution c. Systems of rules governing exchange (redistribution and recognition) d. lifeworld or system of everyday life reproduction	
7. Religious freedom	c. Systems of rules governing exchange (redistribution and recognition) d. lifeworld or system of everyday life reproduction	

Source: developed by the author

contention, conventional analysis of the Ogoni conflict denies any continuity between Ogoni demands and the institutional contexts of its emergence. To address the question, we need to examine the text of the OBR.

The OBR: cultural basis of demands

Ogoni discourse of demand derives directly from the discourse of the nation state as a federal system. The discourse directly addressed existing issues in national politics and the federal constitution. The discourse did not emerge from nowhere; rather, it represents a self-critical emanation of Nigeria's political culture. There is no word like 'exclusion' in the demand. The Ogoni did not see themselves as 'excluded', and therefore fighting to be included. Rather, their discourse challenged the terms of their inclusion and queried those terms as embodying a relationship of domination, which was in violation of the values of Nigeria's federalism.

While environmental pollution and political marginalisation are common-place, conflict is not. There is, therefore, the need to understand why conflict emerges where it does. A consideration of the general conditions in their imbrications with place-specific characteristics is called for. The intersection of the general and specific creates the conditions for emergence of the conflicts.

Much of the data presented here are available in the existing literature. However, they have hardly been considered within the framework of place, which asserts that conflict is inherently geographical, that is, structured by the core elements of place: locale, location and sense of place. Thus, the chapter argues that insights from the resource availability and society-rooted literature provide the precipitating backdrop to the conflict. Place-specific factors, notably the history of struggle and environmental disaster, sense of attachment to place, existing religious worldviews and place-informed framing combined with broad societal conditions, precipitating the emergence of the Ogoni movement.

6 From grievances to micro-mobilisation
How the Ogoni mobilised

Introduction

A slow but steady transformation of Ogoni from a quiescent rural oil-bearing community to a field of confrontation between repressive authorities and the Ogoni took an overt dimension in January 1993. Precisely, on 4 January 1993, an estimated 300,000 Ogoni, under the aegis of the Movement for the Survival of Ogoni People (MOSOP), embarked on an unprecedented protest march against the Nigerian State and Shell Oil Company (MOSOP 2004). The success of the march and latter internationalisation of the struggle elicited repressive measures, which culminated in the hanging of Ken Saro-Wiwa and eight other Ogoni leaders on 10 November 1995. Prior to the march, the Ogoni had presented the Ogoni Bill of Rights, which articulated Ogoni demands, to the central government in October 1990. On 3 November 1992, the MOSOP issued a 30-day ultimatum to the oil companies, namely Shell, Chevron and the Nigerian National Petroleum Corporation, which demanded that the Ogoni should pay royalties and compensation for the devastation of Ogoni or quit (MOSOP 2004). Expressive of the repressive bent of both the State and oil companies, the bill and ultimatum were ignored.

Instead, worried about the effect the Ogoni protest might have on other aggrieved communities in the region, the State attempted to compromise the Ogoni leadership. The Governor of Rivers state held several meetings with Saro-Wiwa in an effort to persuade him to drop the burgeoning struggle. On four occasions (January 1993, February 1993, May 1993 and September 1993) Ogoni leaders were invited to meet with top functionaries of the central government. At the May 1993 meeting, the leaders were asked to identify development projects that they wanted the government to site in Ogoni (MOSOP 2004). The Ogoni leaders, however, rationalised that the mobilisation was beyond mere reform or the conventional practice of buying quiescence through development projects. The goal was to effect a change in the nature of relations between the Ogoni and the State and the former and Shell (Saro-Wiwa 1995). Seeing that its antics had failed, the State adopted a hardline posture, including repression of Ogoni leaders. In April 1993, the government promulgated the Treason and Treasonable Offenses Decree, which made the demand for any form of political autonomy a capital crime. The decree aimed at combating Saro-Wiwa and MOSOP (Ogoni Charities 2012). On its part, Shell embarked on a campaign of calumny against Ken Saro Wiwa, arguing that MOSOP did not represent the entire Ogoni people, and accused Saro-Wiwa of a secessionist agenda (MOSOP 2004). Throughout 1993, Saro-Wiwa was subjected to a spate of arrests and detention (MOSOP 2004).

On 2 June 1993, the Steering Committee of MOSOP unanimously voted to boycott the 12 June 1993 Presidential elections, a position that the Ogoni adopted (MOSOP 2004). In reaction, Ogoni leaders, Dr G.B. Leton and Chief Edward Kobani, resigned their positions as President and Vice-President of MOSOP, respectively. While in detention, Saro-Wiwa was elected President in absentia by the Steering Committee. Suddenly and without provocation, first, the Andoni (June–September 1993), followed by the Okrikans (August 1993), and the Ndoki-Ibo (April 1994) embarked on armed attacks against the Ogoni using very sophisticated weapons. It is widely believed that the attacks were state sponsored. On 21 May 1994, in controversial circumstances, four pro-government Ogoni chiefs were murdered during a meeting at the Palace of the Gbenemene in Giokoo, Gokana. The event provided a pretext for the militarisation of Ogoni, arrest and detention of Saro-Wiwa and many other leaders of MOSOP. The 'wasting operations' of the military left in their trail hundreds of men, women and children maimed or dead (HRW 1995). The military tribunal set up to try Ogoni leaders found Saro-Wiwa and eight others guilty of the murder of the four chiefs and sentenced them to death by hanging on 10 November 1995.

The mobilisation that assumed overt complexion on 4 January emerged more from the framing activities of MOSOP leaders than structural causes. The success of Ogoni framing was not a matter of course; they emerged through experimentation and failures. Activists who were interviewed show that they attended meetings addressed by Saro-Wiwa, and other activists, on invitation by friends or relations. It was at such meetings that various frames were deployed, reconstructed, abandoned and learned. Structural conditions of exploitation and marginalisation were reinterpreted in ways that sought to clarify to the ordinary Ogoni the link between his/her everyday life experiences in the farms and waterways and the action and inaction of the State and Shell. Such grassroots-based reinterpretation marks a difference between the decades of exploitation and quiescence, and the 1990s when the Ogoni danced their anger on the streets.

A major problem with the rich literature on the emergence of the conflict is its focus on meta-narratives that attribute causality to macrostructures. By focusing on macrostructures, meta-narratives hide the everyday activities of actors that galvanised the Ogoni conflict. We cannot fully understand why and how the conflict emerged without considering the purposeful and interpretive actions of Ogoni actors. Moreover, meta-narratives cannot explain why some Niger Delta communities with experiences similar to the Ogoni have not mobilised. Indeed, Eghosa Osaghae asked the pertinent question why the relatively better off Ogoni mobilised against the State and not any other community.

Attempts to explain the Ogoni conflict include the following accounts. Obi (1997) embeds the conflict squarely in the processes of globalisation, and Osaghae in the National Question (Osaghae 1995a). Ikelegbe (2001) attributes the conflict to neglect, impoverishment and marginalisation, which has generated anger, frustration and hostility. Okonta (2008: 5) similarly claims that minority marginalisation, local elite anger and rural dispossession 'exploded' into Ogoni militant nationalism in the 1990s. Thus, the authors suggest that grievances arising from the experience of exploitation and marginalisation compelled the Ogoni mobilisation. While grievances are ubiquitous, conflicts are not. The practical steps taken by these activists to mobilise action is given little thrift in the accounts.

Existing accounts give too little attention to why individual Ogoni participated, or did not participate, in the movement. Such analysis suggests that members of the movement automatically participated in reaction to structural conditions or awareness of dispossession and that they exhibited similar levels of commitment. Moreover, the role of frames in recruitment are minimised. This chapter transcends meta-narratives by focusing on these ideational factors and micro-mobilisation activities of activists to understand the emergence of the conflict. A fuller account of the emergence of the Ogoni conflict needs the benefit of what the Ogoni actors did to move from conditions of grievances to action. This chapter endeavours to shed some light on these activities.

The process of identity formation is central to the ability of mobilised communities to articulate claims and exercise power in society. There is a need to understand Ogoni collective identity making. To describe the process of forming collective identities, we may see it as the making of Ogoni identity, as different from any idea of preformed, pre-existing Ogoniness. This is suggestive of a process-oriented approach. It involves the process of creating shared meanings and consciousness among diverse individuals within the Ogoni cultural space, and the framing of grievances.

If one rejects the idea of primordial identities and assumes that individuals have multiple identities and loyalties, then the question of how the individual belief system is conjoined to the larger goals of collectivity is a key component in a process of forming identity. Groups do not merely define themselves in opposition to others but also they frame their visions of citizenship and narratives of a just society. At the base of the participatory citizenship frame is a belief that Ogoni has been denied participatory rights in a supposed democratic state. The Ogoni could not realise their God-given potentials and fulfil their roles as citizens because they are denied representation in political and economic institutions of the State.

This chapter employs framing as an analytical tool to explore the interpretive work activists executed in the process of mobilising contention. It seeks to highlight the self-directed or local frames, without ignoring the global frames, the Ogoni employed to galvanise mobilisation. As a result, it suggests that theoretical attention to both local and global, and otherworldly frames used by the movement provide deeper understanding of the role of framing in the conflict.

The uses of framing

A frame refers to an 'interpretive schemata that simplifies and condenses the "world out there" by selectively punctuating and encoding objects, situations, events, experiences and sequences of actions within one's present or past environment' (Snow and Benford 1992: 137). Collective identity formation involves framing processes that generate collective action frames, or identification of a problem, which requires collective action, and collective identity frames, or alignment of individual and collective identities. The 'collective action frames are action-oriented sets of beliefs and meanings that inspire and legitimate the activities and campaigns of a social movement organization' (Benford and Snow 2000:614). Framing theory states that activists employ familiar interpretative schemas to attach meaning to events and experiences in order to inspire and legitimate contentious action (Benford and Snow 2000; Snow et al. 1986; Snow and Benford 1988).

Frames perform three functions: diagnosis of the social condition in need of remedy; prognosis on how to actualise such a remedy; and rationale for action. Frames do matter but are insufficient by themselves to mobilise people. For frames to be effective, certain conditions must occur. First, representation of the other as the enemy must concretise the enemy. Mobilisation must be against a physically present, known enemy or, one whose interests are present. Through rhetoric, leaders must provide social actors courage to redefine reality, force change, and to endure throughout the campaign (Lachmann and Pichardo 1994). Benford (1993) argues that simply because people agree with a diagnostic frame on the existence of a problem is no guarantee that they will give up their own endeavours and help alleviate the problem. There is a difference between identifying a problem and convincing people that the problem is serious and deserving of urgent action. Snow and Benford (1988) hold that prognostic framing advances a solution to the situation defined as a problem.

Benford (1993) holds that a sense of moral duty is essential to action mobilisation. Thus, the closer a frame is to providing solutions to a problem and implementation of the solutions, the higher the mobilising capacity of the frame (Gerhards and Rucht 1992). Movements must be involved in the motivational task of constructing and amplifying beliefs about why it is proper and moral to take action to ameliorate a situation. Benford observes that without such framing, micro-mobilisation efforts might stay stuck in the consensus mobilisation stage.

Framing the Ogoni movement

MOSOP's cognitive interpretive activities geared at forming Ogoni collective identity did not emerge in a vacuum, and were not unrelated to cultural traditions and action repertoires. First, they merely wanted all Ogoni under a unified administrative framework. Then they wanted their own Native Authority. Second, they wanted Rivers state in order to escape Igbo domination. Third, they wanted rights as equal citizens of Nigeria.

Framing deals with how actors read and define a situation, apportion blame and advance arguments for change. Framing is an interpretive exercise or worldview that serves to mobilise otherwise quiescent groups. A significant aspect of the literature largely ignores this dimension of the Ogoni conflict. Instead, many scholars have focused on how poverty, pollution and political marginalisation angered the people who consequently resorted to collective action. Missing in the latter explanation is the interpretive works that preceded collective action.

A few scholars have, however, explored the interpretive dimensions of the Ogoni movement. Ogoni identity was 'discursively and politically produced', and rested on 'the invention and reinvention of tradition' as there was 'no pan-Ogoni myth of origin' (Watts 2004: 69) and, Ogoni identity is a recent construction (Isumonah 2004). How insurgents market themselves matters in whether they secure global support. Movements 'magnify their appeal by framing parochial demands, provincial conflicts, and particularistic identities to match the interests and agendas of distant audiences' (Bob 2005: 4). Social movements in the Niger Delta adopted global discourses in the effort to secure international support (Obi 2009). Bob (2005) argues that insurgent groups frame their cause to match the interests of external actors in the quest for survival and growth. Thus, the Ogoni framed their protest as an indigenous peoples' struggle. That explains why the MOSOP variously reframed its goal in ethnic, genocide and environmental terms (Bob 2002).

Clearly, existing literature on Ogoni framing works has given overwhelming attention to the external orientation and roots of Ogoni frames. What is, however, missing is the locally informed and oriented frames that developed in tandem with global external frames. Neglect of the local dimension of Ogoni framing suggests erroneously that Ogoni mobilisation was wholly a function of external resources. The chapter seeks to fill the gap and contribute to the literature by highlighting the often neglect interpretive work that shaped the Ogoni movement. In the context of the Ogoni, the author argues that three frames; oppressive order master frame, miideekor frame, and otherwordly frame, provide insight into how the Ogoni uniquely combined self and other directed frames to mobilise themselves and external support simultaneously.

Oppressive order master frame (OOM)

From the Ogoni Bill of Rights (OBR) and other publications by Ogoni actors, we can derive what may be termed the OOM. Frame refers to 'the conscious strategic efforts by groups of people to fashion shared understandings of their world and of themselves that legitimate and motivate collective action' (McAdam et al. 1996: 6). MOSOP achieved the objective of fashioning shared understanding and motivating collective participation by deploying various frames that aligned its struggle with global discourses and symbols. For instance, MOSOP timed its protest march to coincide with the UN celebration of World Indigenous Peoples' Day on 4 January 1993. MOSOP deployed frames of genocide, ethnic minority, environment and indigenous rights and actively built trans-local networks of connections with international organisations concerned with those issues. Thus, Obi (2009) correctly observes that the movement adopted global discourses in the effort to secure international support. Ogoni activists emphasise marginalisation and exploitation in articulation of why they mobilised (MOSOP 2004). Below are some structural elements of an OOM evident in Ogoni publications (compiled by the author).

- The elites of the majority ethnic groups, their clients from minority groups, the State and Shell compose a federal system, a colonising order.
- The colonising order serves the interests of those who compose it while exploiting and marginalising the minority oil-bearing communities.
- The root of the exploitative nature of the federal system lies in its productive, appropriative and distributive systems.
- The system of exploitation has given rise to numerous problems for oil-bearing communities, including impoverishment, land confiscation, environmental destruction and political marginalisation.
- MOSOP aims to address these problems by attacking it at the root, by restructuring the structure and administrative workings of the federal system.
- The Ogoni outline their strategy as one of grassroots mobilisation, local demonstration and appeal for the intervention of the international community.

In the OOM, the dominant portrayal of Ogoni is as space controlled and exploited by the State and Shell in ways that ignore the environmental degradation and wellbeing of its inhabitants. The nature of rule enriches places and actors beyond Ogoni at the expense of the latter. The Ogoni need to re-establish control over the environment, benefit from the resources therein and, overturn colonial exploitation. The OOM provides numerous entry points for other groups. The prognosis, which emphasises skewed federalism and the need for a restructuring is shared by a broad section of elites and most ethnic groups in the country. The discourse of environmental protection and human rights found resonance with actors at the national and international levels. Prior to Ogoni framing activities, the oppression of the Ogoni was not evident to the human rights and environmental communities. Bob demonstrates that it was the persuasiveness and resonance of the frame that engendered the perception and acceptance by various global actors that Ogoni was indeed a victim of human rights abuses and environmental degradation (2002).

Miideekor's (property owner rights) frame (MF)

Interviewees claim that MOSOP had attempted grassroots mobilisation through the use of global frames of 'indigenous people', 'ethnic minority', 'true federalism' and 'human rights' to no avail as the people could not relate concretely with these frames. To the contrary, the miideekor frame resonated instantly and to the extent that the one-word frame was translated into a song that was sung during rallies held by MOSOP (Chief Emmanuel Nkalaa 2012 int.; Rev. Richard Biragbara 2012 int.). The frame aided Ogoni mobilisation because it enabled everyone, literate or illiterate, to understand the rationale behind the protest, and why everyone participated in the struggle. Also, it lay bare the degree of exploitation of the Ogoni and the depth of Shell's indebtedness to the Ogoni (ibid.; Legborsi Saro Pyagbara 2012 int.; Wilfred Tanee 2012 int.).

Miideekor is an Ogoni word that refers to the palm wine produced in one out of the five workdays of the Ogoni work week, namely Deemua, Deebom, Deezia, Deezion and Deekor. Traditionally, the palm wine tapper may keep the palm wine produced in four out of the five days. However, as Carolyn Nagbo emphasises, the remaining day's production belongs to the property owner (2008 int.). Deekor (or the first day of the week) was a special day for showing appreciation to the landlord. The process of returning the one day (Deekor) a week proceeds due the property owner is miideekor. Miideekor symbolises the relationship between the owner of a palm field and the palm wine tapper, which is a powerful cultural symbol for the Ogoni. An ordinary Ogoni woman, the late Rhoda Komdu Nwinaalee, at a meeting of the Ogoni in Finininane Suanu, Nwike Conference Hall in Bori, where MOSOP leaders struggled conceptually to explain the rationale of the movement, retrieved miideekor from the Ogoni cultural repertoire (Alonale-Laka 2002). One can distil elements of what may be seen as the MF from informants' articulation of grievances and activists' publications (compiled by the author):

- The oil resources in Ogoni belong to the Ogoni.
- It is only right that as property owner, the Ogoni receive a fair share of the resources.
- However, the State and Shell who are tenants on Ogoni land conspired to deny the Ogoni their due.
- At the same time, Shell pollutes and confiscates Ogoni land without compensation, and with impunity.
- This situation is thievery, exploitative and unjust.
- What the Ogoni demand is a fair share or miideekor; not everything.

Applying this cultural frame, the Ogoni defined themselves as owners of the oil in Ogoni and the State and Shell as the tenants. What they expected from the latter is their miideekor or a fair share as property owner. In the cognitive frame, Damgbor Moses argues, miideekor is a widely shared vocabulary in Ogoni (2008 int.). Its deployment in the struggle served to construe the State and Shell as thieves, exploiters and oppressors who deny the Ogoni their right. Moses argues that because miideekor resides in everyday experiences, its frame was coherent and resonated with all Ogoni, greatly aiding mobilisation (ibid.). The next section provides evidence of how activists ingeniously weaved a moral otherworldly basis or legitimacy for the OOF and MF.

Otherwordly frame (OF)

A group of warriors and spirit mediums founded Ogoni and shrines are dedicated to these founding ancestors. Reverence for the ancestors with their warrior ethos and for the power of the spirit mediums has dominated Ogoni belief. The Ogoni believe in a supreme goddess, Bari, who observes events on earth from the sky, and worship ancestral spirits of Ogoni. Beliefs in both the Christian God and Ogoni deities provided moral incentives to contention. Saro-Wiwa linked the beginning of his activism to 'the Voice', which commanded him to work for the liberation of the Ogoni and all other oppressed peoples in Nigeria (MOSOP 2004).

> One night in late 1989, as I sat in my study working on a new book, I received a call to put myself, my abilities, my resources, so carefully nurtured over the years, at the feet of the Ogoni people and similar dispossessed, dispirited and disappearing peoples in Nigeria and elsewhere.
>
> (Maier 2000: 93)

There was an overwhelming belief that the Christian God and Ogoni ancestors were involved in the mobilisation, and that Saro-Wiwa was divinely appointed to liberate the Ogoni (Tonwe-Kpone 2008 int.). Christian churches framed the Ogoni problem in Biblical terms and provided sanction and spiritual support for the struggle by invoking divine intervention on behalf of the Ogoni. Interviewees acclaimed that prayers to the gods gave them the confidence and hope of victory. God and Ogoni deities were involved. According to Chief Emmanuel Nkalaa of Kaani 11 Community, a retired civil servant, 'I saw before my face how the gods came out in full to deliberate on our problems and measures that should be taken to address them' (interview 22 May 2012, PH).

In the Ogoni worldview, a Nwiayor is a juju (idol or deity) priest, and a person who protects is considered a god. Therefore, the god who comes in human form is a Nwiayor. Saro-Wiwa exhibited these attributes. Interviewees assent that they believed Saro-Wiwa was sent by God to his people. Before the struggle began, God, deities, even the spirit of Paul Birabi (foremost Ogoni nationalist leader) and other departed heroes were consulted. Ogoni Council of Churches held prayer meetings for the struggle. For the first time in my life I saw various juju priests I have heard about but never seen when they came out to hold a meeting (Chief Nkalaa, int.). They informed us that the gods were behind the struggle and have agreed for them to contribute to successful prosecution of the struggle, and that victory was certain. Our consultation with the Ogoni Council of Churches showed us that our cause was a just one. It was just in the sense that if one fights for what is right or a good cause, God and the gods will support.

We have been involved in various battles in the past and our gods supported and gave us victory on every occasion. The traditional native doctors besides telling us the mind of the gods, made good luck charms for our protection. We saw these charms work on battlefields before as they protected our youths from dangerous weapons. Bullets could not penetrate the bodies of the youths (Young Kigbara 2012 int.). Activists claim that before the January 1993 protest, groups of Ogoni consulted the Christian God, ancestors, Ogoni deities, and Ogoni spirit on the struggle, and there was a common otherwordly consensus that the struggle was legitimate and that the Ogoni would be successful.

The power of the frames of divine commission suffered a setback when cracks emerged within MOSOP. Sensing an implosion, Kpone-Tonwe attempted to solidify the frame by emphasising otherworldly frames of unity and forgiveness.

> I told the house [gathering of Ogoni elders] that what was happening in Ogoniland was a revolution; that in history, a revolution is caused by a 'spirit force', which rests on a single individual; that in the case of Ogoni, that individual was Ken Saro-Wiwa; that in a traditional setting, based on my study of Ogoni tradition, what the elders used to do, was to find out the individual on whom this 'spirit force' rests. Once that individual has been identified, all the elders used to rally round that individual to give him their support, while at the same time sinking their personal differences or disagreements. I appealed to them to do the same with Saro-Wiwa.
>
> (Kpone-Tonwe 2003: 63)

During personal interviews, the author prodded Kpone-Tonwe on his ideas of 'revolution' and 'spirit force'. He elaborated that as a son of a prominent traditional chief in Khana Local Government Area, he observed from youth that the Ogoni usually consulted the oracles whenever they went to war.

> The consultation is based on the belief that there is a spirit of victory or revolution. Thus, the war is led not by the strongest or most senior general, but one that would lead the army to victory. Once he is identified by the oracle, he is made a leader and supported by all. Based on my sense of history and observation of the Ogoni struggle, the spirit of revolution was on Saro-Wiwa, and, traditionally, all Ogoni supported such a leader regardless of existing differences because he is seen as divinely chosen
>
> (Kpone-Tonwe 2008, personal interview)

The OOM emphasised macrostructures, federalism and the State/Shell alliance, as the bane of Ogoni despite attributing blame to specific actors. As such, it depicted the Ogoni as powerless actors struggling against powerful forces. While it attracted global sympathy and support, the frame had less value for self-mobilisation because it made little sense to ordinary Ogoni (Bari-ara 2006; Legborsi Saro Pyagbara 2006, 2007 int.). The weakness of the OOM and problem of mobilisation was resolved when Nwinaalee reframed the struggle in cultural terms. Miideekor recast the struggle as between a property owner (Ogoni) and tenant (the State and Shell), thus, refocusing attention on present actors and reversing the power relationships inherent in the OOM. It riveted attention on everyday relationships between the Ogoni and Shell, rather than macrostructures. MF, therefore, provided a vocabulary of action, particularly in the event of conflict, rooted in cultural antecedent. The breach of miideekor expressed a breach of cultural expectation and trust, mobilising the Ogoni emotions in ways that OOM did not.

The OF drew exclusively on the Ogoni religious belief systems. Although the belief system included Christianity, a global religion, the use of biblical frames was hardly geared at winning the support of the global society of Christians. The religious frames, drawn from Ogoni worldviews established in a way that it could not to the non-Ogoni, that the mobilisation was legitimate, required to overthrow structural oppression and bound to succeed. Moreover, through dynamic otherwordly interactions and communications between supra-human beings and the Ogoni, the latter were assured of the present help of otherwordly beings in the struggle. The OF facilitated the transformation of the Ogoni from an oppressed powerless actor to a landlord that could exercise power in relation to the oppressor, the point of the MF. With the frame of otherworldly support very well established, and based on historical memory of divine support, the Ogoni had little inhibition not to participate in the struggle.

Far from the strategic bent of clever leaders, an ordinary woman deployed the MF, thus affirming the argument by Aminzade and Perry (2001) that framing can be the unwitting recourse to deep-seated cultural influences. Moreover, contrary to Bob's (2005) wholesale depiction of Ogoni frames in instrumental terms, the MF and OF are no less about control of the modality of being as they are about material benefits (Agbonifo 2009). The redefinition of macrostructures in the grammar of property owner and tenant, provision of moral or otherworldly sanction and mobilisation of passion allowed for the resonance of the MF and OF among the Ogoni. Without the frame of overt support of otherwordly beings for the cause, it would have been a gargantuan task getting the deeply religious Ogoni to embrace the struggle. Together, the OOM, MF and OF frames provided the interpretive lens and basis for mobilisation of material and symbolic resources that galvanised the struggle.

Ogoni framing activities and action motivation

Diagnostic framing

Collective identities traditionally have been established around class (elites and non elites) and cultural categories (Kingdoms). These identities create and sustain group boundaries and are oriented to particular group-specific goals and interests. The Ogoni have also been divided along political party lines during electoral politics. Indeed, much earlier during the Nigerian civil war, some Ogoni elites supported Biafra secession and others the central government. By contrast, the Ogoni movement developed an inclusive collective identity frame that glided over pre-existing categories and divisions. In the new frame, the Ogoni, regardless of status and local elites' domination, were constructed as the poor, oppressed, exploited, and marginalised people as distinct from the rich and dominant ruling elites and Shell. The Ogoni identified the State, Nigeria federalism and Shell as the common problem and the broad solution was ethnic equality that would allow the Ogoni exercise some control over their environment and how their resources are used.

MOSOP employed oppositional framing, arguing that the Ogoni are involved in two grim wars of political tyranny and environmental degradation mounted by both the State and Shell. That way the Ogoni sought to organise different groups around the defeat of a common threat, a colonising and corrupt state and Shell, thereby downplaying the initial difference, and sometimes conflicting interests, among the Ogoni. For instance, it has been argued that there is no pan Ogoni identity, and that at various times in the past the Eleme have defined themselves as non-Ogoni (Isumunah 2004). Yet, during the mobilisation, Eleme and the other kingdoms were mobilised around a common enemy, which helped in downplaying any notion of difference. Participants constructed the broad collective identity of the Ogoni to replace that of the specific kingdom, elites and commoners. The oppositional frame of an 'us' against 'them' emphasises the meaningful unity of the in-group members of MOSOP.

The Ogoni Bill of Rights (OBR) preamble contains a concise statement of the ecological and political situation of the Ogoni. It identifies the condition as problematic, attributes blame for the problem to concrete actors and shows why ameliorative action is required. The diagnostic framing in the OBR may seem radical but it would be a mistake to attribute it to the radical nature of the Ogoni leaders. What changed with the emergence of overt collective mobilisation in 1993 was not the Ogoni leaders or the structural condition of environmental degradation, but the interactional context that facilitated a representation and reinterpretation of existing conditions as bad and unacceptable (Saro-Wiwa 1995: 147–148).

MOSOP also utilised interpretive frames that were meaningful to the large pool of ordinary churchgoing Ogoni to flesh out a common identity, and who the enemy was. Christian religious leaders throughout Ogoni helped in defining the Ogoni situation by likening it to the biblical story of the enslavement of the Israelites in Egypt (ibid.: 120–1). In the religious metaphor, the Ogoni assumed the identity of the oppressed Israelites in contradistinction to the enemy identity of the oppressive Egypt or Nigerian State. The Ogoni religious belief system became a tool for questioning the role of the State and Shell in Ogoni.

Prognostic framing

Snow and Benford (1988) hold that prognostic framing advances a solution to the situation defined as a problem. MOSOP deploys various prognostic framings in its communications. Saro-Wiwa argues that the nemesis of Nigeria inheres in its political administrative structuring: 'As organized today, the country is not a workable possibility. There is no country. There is only organized brigandage' (1992: 91). He points out that aside from political structuring and the inequities of revenue allocation, the administration of the country works against the Ogoni because the wielders of power – the ethnic majorities – administer by cheating.

At the annual luncheon of Kagote on 26 December 1990, Saro-Wiwa outlined a strategy for realising their goals. He argued that the Ogoni have an agenda to which every Ogoni man and woman must commit.

> Wherever an Ogoni man or woman may be, he must not forget our agenda to save our nationality, our language, our culture, our heritage. Ogoni people must co-operate with one another, as individuals, as groups, because that is the only way we can survive. Wherever they may be, they must proclaim their Ogoniness…I believe that the Ogoni agenda is the only one that can save Nigeria from future destruction. This agenda postulates the equality of all ethnic groups, big or small, within the Nigerian federation as well as the evolution of proper, undiluted federalism in the nation.
>
> (Saro-Wiwa 1995: 75–76)

The Ogoni master frame went beyond diagnosis and prognosis to suggest means for reaching the proposed solution. The solution to the Ogoni problem rests on granting the Ogoni people political autonomy to participate in the affairs of the Republic as a distinct and separate unit by whatever name and equal treatment of the component ethnic groups regardless of size.

Employing a culturally resonant frame, MOSOP argue that resolution of the Ogoni conflict must be based on the granting of the Ogoni its miideekor. The miideekor frame clarifies to the ordinary Ogoni that what MOSOP demands, far from being impossible, is a minute fraction of what Shell and the State extract from Ogoni. Payment of miideekor was a simple cultural expectation of a tenant. Moreover, the frame recentralises the Ogoni identity as owner of the land, which confers on the latter the cultural right to drive away an ungrateful thieving tenant, Shell.

Motivational framing

In contrast to oppositional frames, prefigurative frames represent articulations of what MOSOP stood for, not against. Like prognostic frames, prefigurative frames proffer a common vision for the future, but lack the specificity of prognostic frames. MOSOP evinced a vision of society characterised by ethnic equality, social justice, participatory democracy and transparency. Prefigurative politics refers to the participatory democratic bent of movements and the latter's attempt to embody personal and anti-hierarchical values in politics.

Motivational framing involves a call and rationale for, involvement in ameliorative action. Activists need to create motives for participating in a given action. To Saro-Wiwa, the exploitation of Ogoni amounts to theft, slavery and genocide. Environmental motivations arose when the Ogoni argued that destruction of their environment, the theft of their lands and resources, would lead to their extinction. The risk of extinction and genocide proved useful motivation for the Ogoni to take immediate action to avert their own extinction. Movement leaders encouraged members not to be discouraged by the odds because the Ogoni can and must do something about it and that the responsibility lies with every single Ogoni (Saro-Wiwa 1995). During the march on 4 January, Kobani encouraged the Ogoni to be fearless because they would not be harmed on their God-given land (ibid.).

Saro-Wiwa described Ogoni collective action as a key civic duty (MOSOP 2004: 50). By appealing to the pre-colonial autonomy of Ogoni, the need to reject internal colonialism and re-establish the status quo ante, Saro-Wiwa deployed moral incentive. When Saro-Wiwa (1995: 74) used the spectre of hopelessness for Ogoni children and gradual extinction and, exhorted the Ogoni to act to save themselves, he used negative incentive as well.

While MOSOP appealed to global frames of human rights and international support to motivate collective action, MOSOP itself had roots and motivation in the Ogoni cultural tradition of inclusive organisation. In 1992, a committee of Ogoni scholars recommended that the MOSOP look inward and draw upon ancient cultural resilience to be effective (Kpone-Tonwe 2003: 75). Saro-Wiwa warmly embraced and implemented it, touring and mobilising all segments of Ogoni society. Cultural antecedents informed Saro-Wiwa's grassroots strategy. By the principles of Yaa, the traditional means by which young Ogoni men train for leadership and life, and Ogoni elders identified and subtly recruited talented youths into secret societies, the Ogoni successfully organised their community maintaining their autonomy and living with pride among their more powerful and less powerful neighbours (Kpone-Tonwe 2003).

That fact alone suggests that MOSOP's prefigurative politics included societal organisation in which everyone was useful and had a role to play. Ogoni deployment of participatory democracy and social justice, which are arguably global frames, was not entirely informed by the latter. Making the movement an inclusive one created space for the excluded young people and women to have a voice in matters that affected them. Without such social empowerment it would have been difficult for MOSOP to elicit the widespread support it achieved.

MOSOP's prefigurative politics referenced a mythical communal past when life was Edenic, and there was little or no suffering. Consonant with this vision MOSOP's desired society was one characterised by egalitarianism and downwardly accountable social institutions. Adler (2012) argues that the mythic roots of prefigured society notwithstanding, it has capacity to strengthen hope spawned by a movement. As a specific form of motivational framing, prefigurative framing becomes a strategy to excite the collective imagination and galvanise people around a vision of the desired society.

Apart from providing a rationale for collective action, movement activists also framed a vocabulary of motive, which persuaded recruits and potential recruits that collective action would produce desired changes. The Ogoni worldview retains a subsisting belief in a mythical being called Nwiayor whose advent will mean the Ogoni liberation. To Kpone-Tonwe, the Ogoni historian, pastor and university professor, the spirit of revolution rests on such an individual and the Ogoni tradition dictates that they rally around that person (2008 int.). The idea is that, because this person is the chosen one, they succeed. When Saro-Wiwa appeared with his gospel of freedom, the people eagerly embraced him as one sent by God to liberate the Ogoni. This belief was a compelling reason for mass recruitment into mobilisation (Tonwe-Kpone 2008 int; Agbonifo 2003; Tschirgi 2007: 125).

The belief in the Nwiayor and leadership of the ancestors are firmly rooted to Ogoni as 'place'. The Nwiayor was to descend to a specific place, and the ancestors inhabit specific places. To the extent that such cultural cognition served to mobilise the Ogoni, place became important. MOSOP drew upon this sacred repertoire. The symbol of the Wiayor and the narrative of liberation provided religious assets that mobilised and sustained Ogoni commitment to the struggle. The Ogoni believe God sent Saro-Wiwa to liberate them; thus, the people took whatever he said as divine and final (Kpone-Tonwe 2008 int.). Even Kpone-Tonwe admitted such when he told a group of Ogoni chiefs that 'the spirit of revolution rested on Saro-Wiwa' (ibid.).

These beliefs turned Ogoni into a space of emotions, a complex of natural and supernatural forces so that in the face of all odds, the people believed nothing could hurt them on their God-given land and that their cause would triumph. Interviewees explain that the Ogoni believed that their ancestors not only supported but also led the struggle. 'To be honest with you, every Ogoni believes that there is a spiritual touch to everything. The belief in ancestral leadership is a culture of Ogoni' (Bari-ara Kpalap 2008 int.). According to Chief Emmanuel Nkalaa 'Yes, God was behind the struggle, even most sincerely, the Ogoni gods never slept throughout the whole process' (2012 int.). To the activists, this explains why the Ogoni viewed anyone or anything they see as betraying the cause of the struggle with anger and suspicion.

Conclusion

Little scholarship attempts to understand the interpretive dimension of the Ogoni mobilisation. Existing accounts of the Ogoni conflict, which engage with the meaning works of the movement attribute causal significance to the role global frames played in the emergence of the conflict. While it is an improvement on the overwhelming focus on the causal powers of macro-structures in the literature, the interpretive turn ignore the role of local frames, suggesting that the latter were irrelevant and insignificant. The chapter transcends the glaring limitation of the framing approach to the Ogoni conflict by highlighting the frames, and their roles, that have been obscured in the literature. We cannot fully understand the rise of the Ogoni conflict without paying clear attention to the three layers of frames the Ogoni simultaneously weaved to access global, local and otherwordly support.

The Oppressive Order Master (OOM) frame depicted Ogoni grievances in structural terms. It depicted the Ogoni in global frames as a 'marginalised' powerless 'minority group' at the receiving end of the powerful State and Shell whose actions are to blame for Ogoni environmental and social problems. While the frame enabled mobilisation of external support, it had little value in self-mobilisation. The miideekor frame (MF) redefined the conflict in everyday vocabulary of the Ogoni. The redefinition of macrostructures in the grammar of property owner and tenant, provision of moral sanction and mobilisation of emotion allowed the MF to resonate among the Ogoni. The OF provided negative sanctions for the oppression the Ogoni lived through, and positive sanctions for the Ogoni efforts to undo the oppression. Moreover, the OF provided assurances of the support of otherworldly beings for the success of the struggle against oppression. Together, the OOM, MF and OF provided the interpretive basis for the mobilisation of material and symbolic resources that galvanised the struggle.

7 Cultural basis of mobilisation

Introduction

There is a basic difference in the 'nature of the bonds among men', which reflects the difference between the moral order and the technical order (Redfield 1953: 20). The moral order emphasises 'the organization of human sentiments as to what is right,' and the technical order as that 'which results from mutual usefulness, from deliberate coercion, or from the mere utilization of the same means' (1953: 20–21). Some argue that every technical and material change is associated with 'a corresponding change in the attitudes, the thoughts, the values, the beliefs, and the behavior of the people who are affected by the material change' (Foster 1962: 2–3).

Contrary to the materialist interpretation of history, the moral order was not a passive victim of changes in the technical order; ideas or the moral order could serve as founts of social change (Redfield 1953). Despite evidence to the contrary, the possibility that the moral order can act as an independent variable in cultural change remains controversial (Leis 1964). Even in the social movement literature the causal role of culture in mobilisation is unclear.

Political crisis, or structural change, is often posited as the wellspring of collective action. For instance, cultural challenge becomes a mobilising force during momentous or unsettled times (Swidler 1995), or social instability facilitates contentious challenges (Jensen 1995), or contentious groups are enabled to articulate new modes of being in rejection of existing cultural models during crisis and unstable times (Johnston and Klandermas 1995).

The idea of a clean break between stability and contention, and the duality between social structure and countercultural challenge still shapes culturalist analyses, with the effect that the cultural aspects of political and economic structures and opportunities are obscured (Polletta 2008).

Collective mobilisation is seen as an effect of structural conditions, particularly when they are thrown into disequilibrium. Many scholars draw causal links between social structural conditions and the rise of collective actors. A common thread, which runs through the accounts, is the view that political marginalisation, economic strangulation, environmental degradation, and an unaccountable State explain the emergence of collective actors.

In the accounts, the causal factors that are privileged are mainly acultural. Culture is neither approached as entangled with social structures nor conceptualised as a source of mobilisation. Where culture is given attention, it is analysed as a distinct entity or the background to political and economic factors of significance; not as a causal factor in its own right.

The stable quality of social structures is a function of the active reproduction of meanings, which are never settled but always contested (Polletta 1997). By challenging unsettled meanings or reinterpreting them, cultural challenge can destabilise seemingly stable structural arrangements. Cultural challenge is inherent in the process of 'active reproduction of meanings', displaying continuities between existing relation and the challenge that opposes it (Polletta 1997).

While the importance of cultural elements has been recognised, the promise of the recognition has been undermined by the suggestion that unstable conditions foster emergence of cultural challenge. In this view the factor of primary importance is the condition of instability as cultural challenge arises in the context of crisis. Thus, Polletta (1997) argues that if the value of cultural challenge emerges only when the status quo falls in disequilibrium, analytical focus should be riveted on the disequilibrating structural conditions rather than the cultural challenge. However, collective actors do destabilise the social formations that shape everyday life. This vital impact of collective actors has been obscured. How collective actors stagger social structures has received little attention in the literature on the Ogoni because of the secondary importance assigned to cultural challenge in political science texts. Instead, the emphasis has been on acultural structural conditions, which are assumed to invariably mediate cultural challenge.

An added effect of the privileging of acultural social structures has been the tendency to evaluate the significance of collective actors in terms of political and economic resonance. For instance, many have argued that the Ogoni movement failed because it did not achieve its political and economic objectives. In other words, the efficacy of the movement is analysed vis-à-vis the structural conditions that are assumed to have given rise to it. Perhaps, this explains the paucity of scholarly work on the impact of the Ogoni movement. Attribution of primary importance to social structures as a precondition for the emergence of cultural challenge not only obscures how collective actors destabilise structural conditions, but prevents active engagement with a fundamental impact of collective actors.

Acultural explanation of the Ogoni

Okonta (2008) illustrates the foregoing argument. He argues (see below) that because of the presence of older Ogoni elites, the younger Saro-Wiwa had little room to manoeuvre. It is unclear in Okonta's analysis who decides when the room is not enough. Even if we accept methodological individualism as an approach to apprehend Saro-Wiwa, was the 'room' also not 'enough' for collaboration? Was finding 'another political niche' on the federal side the only predetermined outcome of a clogged political space?

Okonta's narrative argues that Saro-Wiwa joined the federal side against Biafra because:

> Ken Saro-Wiwa, we contend, felt politically irrelevant within the new Biafran framework. A historically marginalized Ogoni and a recently marginalized Saro-Wiwa, lacking a political base of his own in Ogoni politics, thus merged in one political entrepreneur. The outcome was a relatively unknown but highly intelligent young man parachuting himself to the top of the Ogoni political pecking order when the civil war ended, and he emerged one of the new men of power in the new Rivers State.
>
> (p.105)

Okonta further contends that 'this tried and tested strategy of breaking rank and finding new political allies in times of crisis served Saro-Wiwa well when the battle for the soul of MOSOP reached its height in 1994' (2008:105). In underlining personal and communal marginalisation as an explanation of Saro-Wiwa's action, the author completely ignores the role of culture in human behaviour. Moreover, he suggests a definition of material human motivation that is totally isolated from cultural influences. Yet, high achievement motivation may be less geared to wealth accumulation than satisfying societal expectation. The Calvinists in Weber's Protestant Ethics pursued capital accumulation not simply because of a materialist aspiration for pecuniary reasons but in order to ensure themselves a place in heaven.

Turning his gaze on Ogoni youth, Okonta explained their mobilisation as a result of the feeling of being betrayed by the moderate stance of Leton, Kobani, Badey and Birabi. The youths were, thus, angry because they were 'desperate for a speedy and effective political settlement of their abject condition' (2008: 222). Okonta suggests that the youths were simply homo economicus whose actions are reducible to bread-and-butter explanation. He continued in the same vein in his understanding of Saro-Wiwa. Saro-Wiwa was a rich hardliner who was independent of the State. Other Ogoni elites were dependent on the State and conscious of the possible consequences of disobedience to the State. But people do not invariably become robots to structures and institutions. The Biafran rebel leader, Odumegwu Ojukwu, was Oxford educated and a seasoned soldier. He knew what military discipline was, yet he rebelled against that institution and the rules of the military in organising a rebellion. Moreover, cases abound of seasoned civil servants who broke institutional rules, enriching themselves corruptly. More recently, former Central Bank Governor, Sanusi Lamido Sanusi, accused the NNPC of sleaze even though he was well aware of what the government could do to him if he squealed. A captain in the Nigerian army recently released an audio recording of a conspiracy to rig the 2014 Ekiti state governorship elections, involving his superior officer, a Major General in the Nigerian Army. The junior officer knew what military discipline means yet he took the risk, not only recording the conspiracy but releasing it publicly. Knowing what the army could do to him, he went underground. Structures and institutions do not invariably explain compliance.

Okonta deploys arbitrary explanatory schemes in his efforts. On the one hand, he reduces Saro-Wiwa and the youths to individual psychological factors, and on the other uses an institutional lens to explain the action or inaction of the moderate elites. Okonta suggests that Saro-Wiwa and the youths lacked institutional embeddedness. They lived above Ogoni cultural traditions and institutions. The civil service was the only state institution that could shape an actor's identity and behaviour. Okonta succeeds at explaining the action of Ogoni youths in material terms, and reduces Saro-Wiwa's motivation to an unexplained radicalism and economic self-sufficiency. For the moderates, they are variously shaped by clientelistic ties to the State, unquestioning obedience to authority and fear of State reprisal. Overall, these critical stakeholders of MOSOP are disembedded from place, from any authentic cultural base. Such materialist framing could only lead Okonta to conclude selfish political interests and intra-elites competition explains the failure of MOSOP.

When Okonta argues that Ogoni politics was nothing more than the competition for personal and provincial advantage, he deploys an impoverished definition of competition. Conflict over material things is not solely materialistic because such conflict involves a questioning of the rules of the game or epitomisation of the conception of what is right. Moreover, competition for personal and provincial advantages does not completely capture the complexity of Nigerian politics.

In a rather confusing and self-contradictory manner, Okonta argues that the Ogoni, including the selfish elites, became '"tribesmen" defending 'tribal' culture to advance their quest for full citizenship, democracy and development' (p.208). Ethnic identity can be revitalised because it is always latent; it never completely passes away. You do not become who you are. The making of Ogoni identity in the early 1990s was not an escape from a Nigerian identity to a tribal one. If at all, Ogoni mobilisation signalled the orchestration of a national identity, deriving as it were from the national value of federalism and equality. It is not clear what that 'tribal' culture is or whose culture it was. Was it the exclusive culture of elites' domination that the moderates represented or the participatory and inclusive culture that MOSOP enunciated? The argument that the Ogoni sought to advance democracy and full citizenship hardly sits well with the argument that Ogoni elites fought among themselves to control MOSOP for personal and clan advantage. He accused Saro-Wiwa, during his struggle with other Ogoni elites, of strategically widening the space of participation for the youth and women in an attempt to control and use MOSOP to outmaneuver other contenders for political authority.

The role of cultural challenge

When does cultural challenge become a mobilising force? Swidler (1995) argues that it is during momentous or unsettled times that cultural ideologies emerge and influence its adherents. Similarly, Jensen (1995) suggests that sociopolitical instability facilitates contentious challenges and penetration of new ideas into established patterns. For Hart (1992), crisis compels individuals and groups to entertain new ways of being and doing things. Johnston and Klandermas (1995) summarise the point when they assert that during crisis and unstable times, contentious groups are enabled to articulate new modes of being in rejection of existing cultural models. The suggestion that instability or disequilibrating conditions enable the emergence of cultural challenge emphasises the primacy of unstable conditions, which spawn cultural challenge. Is it impossible for social movements to catalyse the destabilisation of the institutional logics that inform everyday life? (Polletta 1997: 433).

Many scholars contend that movement participation is driven by narrow self-interests (Chong 1991; Frolich et al. 1971; Olson 1965), and access to resources (Oberschall 1973; Tilly 1978). Many Nigerian scholars have emphasised the role of material and provincial factors, including a political culture of corruption. Less stressed in the academic engagement with the conflict is the understanding that culture can serve to politicise and depoliticise actors for contentious action. The contention here is that cultures generate deeply felt commitments and aspirations, which in turn instigate collective mobilisation. Culture is often portrayed as a mediating variable between structural opportunity and collective action (McAdam 1982). Culture, in effect, has no independent shaping effect on why people decide to participate in collective action (Polleta 1997). Many scholars conceptualise culture in fixed terms (Morris 1984; McAdam 1982; McCarthy and Zald 1977). And as a result, they obscure the interconnection between structure and culture.

Sewell (1992) defines structures as cultural schemas invested with and sustaining resources, which reflect and reproduce unevenly distributed power. The stable quality of social structures is a function of the 'active reproduction of meanings' (Polletta 1997: 434), which are never settled but always contested. The definition not only points at how cultural challenge can destabilise institutional arrangements, but also to 'continuities between existing relations and the challenge that opposes them' (Polletta 1997: 434). Culture may be approached less as people's shared formal worldviews and values than as their ideas about how the organisations and institutions they participate work or should work. Polletta (2008) argues for institutional schemas: models, recipes, and rules of thumb 'underpinning sets of routinised practices around a culturally defined purpose' (p.85). Treating culture as institutional schemas has advantages. One, 'it treats culture as constitutive of interests and identities but also as circulating through networks, backed up by resources, and employed in the service of organizational agendas' (p.85). Once a schema gains dominance, it creates stakes in its enforcement and interpretation and challenge. Second, it helps us to think about mobilisation 'not as the result of long-standing actors with stable interests confronting new political opportunities but, rather as familiar, routinized practices becoming problematic in a way that creates new actors and interests in contention' (p.85).

> The discrediting of old institutional schemas or the ascendance of new ones, conflicts among schemas previously seen as congruent, people's ability to use schemas from one institution as standards for measuring the performance of another institution – each of these developments may generate new lines of contention.
>
> (p.85)

Moreover, seeing culture as institutionalised schemas helps to get at the processes by which culture sets the terms of tactical choice. Familiar ways of doing things and seeing things shape activists' strategic possibilities.

MOSOP and cultural contestation

In many accounts of conflicts in the Niger Delta, political crisis is given as the wellspring of collective action. Collective mobilisation is seen as an effect of structural conditions, particularly when they are thrown into disequilibrium. For instance, many writers have drawn causal links between social structural conditions and the rise of the Ogoni movement. A common thread that runs through the accounts is the view that the longstanding pattern of political marginalisation, economic strangulation, environmental degradation, and an unaccountable State and Shell explains the Ogoni rebellion. Inevitably, such accounts suggest a look along the lines of social structure in order to account for the emergence of MOSOP. In the accounts, causal factors are mainly acultural. Culture is neither approached as entangled with social structures nor conceptualised as a source of mobilisation. Where culture is given attention, it is analysed as a distinct entity or background to political and economic factors of significance; not as a causal factor in its own right. Either way, scholars maintain a dualism of structure and culture with the effect that the mutual imbrication of culture and structure is hidden.

For instance, Okonta casts the Ogoni moderate elites as institutional pawns, and defined Saro-Wiwa and the youths as greedy actors who were not subject to any institutional restraint. Whether subject, or outlaw to institutional restraints, Okonta characterisations of the Ogoni leave no room for consideration of the role of culture in the emergence of the Ogoni struggle and intra-MOSOP conflict. Thus, the author pays no attention to the cultural space of the Ogoni, or the immediate environment within which the movement emerged. Closely related, Okonta fails to consider the Ogoni cultural organisation, KAGOTE, as part of the Ogoni cultural space. Similarly, he ignores the institutional constitutional context, or the nation space, which the Ogoni held responsible for its troubles. What was the impact of these cultural systems on the Ogoni? Prior to the 1990s, were these systems settled and uncontested? Was MOSOP unrelated to these cultural systems or we can trace some continuities between them? Ogoni and Nigeria were sites of cultural contestations. While predating MOSOP, the latter, nevertheless, emerged within the climate of such contestations, and thus stands in continuity with it. It is only by hiding such cultural contestations that personal, provincial and materialist explanations can be foisted on individual actors or collectivity as some scholars do.

A social conflict is characterised by a clear definition of protagonists and antagonists, and the resources or stakes they fight over (Touraine 1985). There are several categories of social conflict, including: the competitive pursuit of collective interests, a political force seeking to change the rules of the game; and, the conflict whose stake is 'the social control of the main cultural patterns, that is, of the patterns through which our relationships with the environment are normatively organized' (ibid.: 754). Constituting the cultural patterns are a model of knowledge, a type of investment and ethical principles through which relationships with the environment are normatively organised. It refers to the conflict between hegemonic deployment of knowledge, investment and ethical principles and the redefinition by the masses of representations of truth, production and morality. Social movements challenge 'the modality of the social use of resources and cultural models' (Touraine 2000: 90). A social movement is never reducible to the defence of the interests of the dominated. Its ambition is always to abolish a relationship of domination, to bring about the triumph of a principle of equality, or to create a new society, which breaks with the old forms of production, management and hierarchy (ibid.: 92). The Subject is manifest by the presence of ethical values that come into conflict with the order of things. A social movement defends 'a social modality of the use of ethical values, and comes into conflict with the modality that its adversary is trying to defend or enforce' (ibid.: 95). However, the ethical dimension cannot be conflated with the discourse of demand. Ethical discourses imply freedom, and basic rights, concepts irreducible to material gain. Touraine further argues that the emergence of the Subject has implications for intra-movement relations and the wider society: forms of intra-group exploitation and marginalisation cannot hide from the Subject.

While the Ogoni movement championed the need for redistributive justice, discursively it left no one in doubt that distributive injustice is located in the architecture of domination. Thus, it equally demanded the abolition of the system of domination, which sustains the minority status and exploitation of oil-bearing minority groups. The Ogoni struggle then gave birth to the Subject, ethical principles, which stood at variance with the ethical values the State and Shell uphold. For instance, the Ogoni argued that Nigerian governments operated a skewed federal structure, which violates the true values of the State. The Ogoni described as redistributive injustice and internal colonialism a system of ethnic relations based on preponderance. They argued to the contrary that all constituting ethnic groups should be treated equally no matter how small or large. The argument for equality and fairness at the federal level put to the test egalitarianism in Ogoni. Just as we must question the democratic credentials of a federal Nigeria, the egalitarian credential of the Ogoni needs to be demonstrated. In the case at hand, the rise of the Subject had implications for intra-group relations among the Ogoni.

The root problem was a skewed federal structure not merely the absence of development projects or patronage. Saro-Wiwa interrogated the very nature of Nigeria's federalism. The feeling of 'exploitation and economic slavery' results from the whittling down of the derivation principle, on which development is based, from 50 per cent to the state of origin to a negligible percentage in the 1980s (MOSOP 2004: 49). He asserts that it is incorrect to argue that such a revenue distribution formula is based on law because the process involved excludes owners of the resources. Denying Ogoni and other minorities such rights makes the 'beloved country a very unequal one or, for the Ogoni and their like, a slave society in which the master groups have all and the slave groups nothing' (MOSOP 2004: 45–46). Osaghae (1995b) argues that 'the Ogoni movement was part of a larger articulation of dissatisfaction with the structure of the Nigerian federation and of power sharing within it by several groups' (MOSOP 2004: 5–6). Osaghae correctly underlines the continuity between the climate of dissatisfaction with the operation of federalism in Nigeria and Ogoni grievances. In effect, the Ogoni challenge to Nigeria's federalism was not emergent and unique to the Ogoni. Rather, it emanates from and resonates with pre-existing widespread gripe with the order of things.

The Ogoni challenge was shaped by the institutions of state, in particular, the federal constitution, which legalises federalism as Nigeria's system of government. In fleshing out the country's federal system, nationalist leader, Chief Obafemi Awolowo, argued persuasively for a federal system that accords equality to its constituting units regardless of size. These were the two fundamental bases for Ogoni grievances with a skewed federal structure and clamour for one that does not discriminate against minorities. Counter-posing the skewed practice of federalism against federalism proper, the Ogoni accused the ruling elites of betraying the true values of the State. Ogoni articulations are shared by a large number of Nigerians, including notable scholars and politicians. Okonta misunderstands the nature of the MOSOP challenge when he argues that the MOSOP and the OBR was the Ogoni attempt to transform conflict of interests 'into a higher order politics by urging Ogoni codes of civic virtue on fellow Nigerian political actors' (p.8). What the Ogoni urged, far from 'Ogoni codes of civic duty', was the faithful practice of a Nigerian constitutional virtue or value – federalism – accusing the ruling elites of betraying that value. They appealed to a national value in an attempt to advance their interest, but more importantly in order to abolish a relationship of domination, which existing federalism represents and reproduces. It is misleading to reduce the Ogoni struggle solely to the defence of the interests of the Ogoni.

In a different way, a form of national institutional practice shaped the Ogoni challenge. The federal government maintained the quiescence and support of factions of the elites through patronage distribution, or 'settlement'. Patronage is usually in the form of political appointment to juicy positions and siting of development projects in the domains of the elites in question. Through these means the State bought elites' loyalty and ensured uninterrupted exploitation of hydrocarbon in the Delta. During the Ogoni mobilisation, the State attempted to compromise Ogoni leaders when a top government functionary asked the former to identify what they needed and stressed the need for projects that could be implemented very quickly. The Ogoni leaders asserted that their demand far outstripped projects. They were aware that development projects, contracts and appointments do not and cannot alter the processes of political marginalisation, environmental degradation, and economic strangulation in Ogoni. Therefore, the appropriate point of attack was the process of dispossession. The demands contained in the Ogoni Bill of Rights reflect the determination to abolish a relationship of domination and dispossession.

Ogoni and KAGOTE

The political dominance of KAGOTE provides an entry into the nature of patterned relationship between, on the one hand, Ogoni elites and the State, and between elites and ordinary Ogoni, on the other. Over decades, several organisations were formed purportedly to advance the interests of the Ogoni vis-à-vis other ethnic groupings in Nigeria. The formation of KAGOTE coincided with the creation of Rivers, and through it Ogoni elites consolidated their status, wealth and power. Through KAGOTE they were able to win lucrative contracts, influence appointment into important positions, and shape political development in Rivers. Moreover, KAGOTE helped to crystallise the common interests of the elites. As power brokers in Ogoni, the elites were enlisted into a clientelist network with the State. As allies of the state in progress, the responsibility fell on them to ensure an uninterrupted flow of crude oil, which is vital to global capitalist accumulation. Dutiful performance of their comprador role ensured their access to wealth accumulation, status, and political power. In this context, the idea of advancing Ogoni interest was replaced with advancement of the elites' interests.

The writ of state runs in Ogoni through a network of ties between the elites and the State. As a consequence of its lucrative links, Ogoni elites transformed the notion of Ogoni gerontocracy, the rule of elders, into a non-participatory system that denied the youths and women voice. The process of modifying the precolonial gerontocratic system was already underway during the colonial era. The culture of quiescence was so pervasive that Shell's oil-related activities polluted farmlands and water bodies at will. Shell claimed ancestral and family land, and destroyed food and tree crops when it choses. Where it decides to pay compensation, it determined the level of compensation with impunity. Understandably, KAGOTE hardly raised a voice of opposition. Where the elders kept mum, of course, no one else could speak; not on behalf of self or others. But that was not all. Local elites allegedly slept with the girlfriends or wives of ordinary Ogoni. Where the need arose they easily could call on their political contact with the police to arrest and intimidate any protesting spouse. Long acquaintance with oppression and the untouchability of the elites helped concretise the culture of silence and deference. They experienced slow death!

How did MOSOP respond to these patterned social relationships between Ogoni elites and the State, and between Ogoni elites and ordinary Ogoni? We cannot read the nature of MOSOP as autonomous and unrelated to broader societal meanings and patterned relations. For to do so is to yank the movement from the context that mediated it. The cultural life of MOSOP must be seen as an extension and expression of ordinary Ogoni cultural life, and not as diametrically different from it in origin and trajectory. Thus, Ogoni pre-existing codes and traditions shape MOSOP, both enabling and constraining it. For instance, Alexander and Smith (1993) argue that opposing sides to political debates in America are constrained to frame their positions in terms of the 'democratic code'. These are standard rules and expectations of the American society. To violate or negate them attracts outright public opprobrium. The MOSOP drew on and resonates with a decades-old tradition of organising, pre-existing cultural codes and structures, and Christian and traditional religious cultures. Also, its frames and strategies were shaped by broader cultural codes, or standard rules and expectations of the Nigerian society, such as democratic participation, equality and federalism.

Some scholars emphasise the external origin or orientation of MOSOP frames (Bob 2002). Such an effort undermines any continuity between MOSOP and the place or culture within which it emerged. Indeed, MOSOP appealed to global frames but it remained rooted in and motivated by institutional traditions in Ogoni and Nigeria as a whole. MOSOP emerged in a cultural context where gerontocracy was the order of the day. Gerontocracy was not simply the rule of the elders; in Ogoni it implied the quiescence of women and youths even in the face of slow violence. It was clear to part of the leadership of MOSOP that the gerontocracy was implicated in the onset of the problem it seeks to confront. This leadership was aware that the elders succeeded in pacifying the Ogoni because they successfully redefined gerontocracy as a system where ordinary people must defer to authorities and remain quiescent even in the face of unbearable pain. To be effective, a committee of Ogoni scholars advised in 1992, MOSOP must be founded on Ogoni ancient cultural traditions, which rested on inclusiveness (Kpone-Tonwe 2003: 75). Through such inclusive traditions Ogoni sustained its prosperity and independence from more numerous neighbours. The assent on inclusiveness was a direct challenge to the exclusive ethos of KAGOTE and the newly formed MOSOP. Against the exclusive ethos of the elders and their wishes, Saro-Wiwa, armed with the committee's report, toured and mobilised all segments of Ogoni society. Based on an understanding of how the Ogoni organised itself in the past, Saro-Wiwa destabilised the rule of the elders and brought the excluded members of Ogoni into membership of MOSOP.

In order to address the question of how the emergence of new movements is enhanced or limited, and how their logic is influenced by broader societal meanings and patterned relations, Polletta focuses on the civil rights movement. Despite the important role of southern Black churches in fostering and sustaining southern civil rights protest, Black ministers were not the avant-garde of the struggle. Many ministers reluctantly joined the cause because they depended financially on Whites. While the livelihoods of ministers in southern cities depended on their parishioners, those of their rural counterparts depended on the Whites. The ministers in the rural areas had to work part time for Whites, making the ministers weak and exposed to pecuniary reprisals (Payne 1995).

Moreover, church leaders had lucrative clientelist ties with Whites. The church leaders who enjoyed such ties were compensated for serving as advocates of only moderate reform. The beneficial pact between the Whites and client ministers was threatened by the emergence of new sets of leaders. In such a compromised position, the ministers could not resolve the tension between maintaining their own interest and the forms of change that would make a meaningful difference in the fortune of their members. Polletta argues that the new leaders displayed willingness to confront, and question the judgment of Black leaders, and criticised the stress placed on 'qualifications' in order to undermine the argument that Blacks without such qualifications lacked capacity for political participation. Importantly, the critique aimed at destroying the existing forms of deference that kept poorer, less well-educated Black citizens from assuming leadership of the movement. The outsiders functioned 'not to empower a powerless group or to enlighten the falsely conscious, but to challenge the structured relations within the group that channeled resistance in an accommodationist direction' (Polletta 1997: 436).

The patterned relations between ordinary Ogoni and their elites, characterised by deference and quiescence from below, gave the elders leeway to channel resistance in an accommodationist mode. The Ogoni issued their Bill of Rights in October 1990. On 3 November 1992, the MOSOP issued an ultimatum to the oil companies and the State demanding compensation for the devastation of Ogoni or quit. On both occasions the Ogoni were ignored and the elites remained too weak to do anything. When the MOSOP steering committee voted to boycott the June 1993 elections, Ogoni elites argued against the position stressing the need not to alienate the State. Also, prior to the 4 January protest march, elements of the elites had shown anxiety over how the State would interpret such action. Together, these suggest the elites were inclined to channel protest in a way that would not alienate the State. However, the power to do so had been undermined with the inclusion of all

Ogoni in the decision-making process of MOSOP. That action ensured that women and youths could deliberate and participate in decisions on issues that affected them. Importantly, the long tradition of deference and quiescence pressed the notion that the youths and women did not have what it takes to make rational decisions. By undermining such a myth, the structured relations that served quiescence was over-turned.

Conclusion: political crisis or culture?

Okonta locates structural conditions as cause of the emergence of Saro-Wiwa and the challenge he represented. Prolonged economic crisis fostered by structural adjustment programme (SAP), authoritarianism of the Babangida military regime engendered old ethnic animosities and created new ones. This coupled with a crisis of legitimacy that enveloped Nigeria forced a retreat from the Nigerian identity to subnational identities; 'Ken Saro-Wiwa emerged in this period of great flux to finally realize his thirty-year-old ambition' (p.169).

The idea of a 'retreat from the Nigerian identity' is rooted in a structuralist understanding of why the Ogoni mobilised. Such an understanding, along with dominant macro-explanations of the conflict, dispense with the cultural basis of mobilisation. The culture of quiescence was pervasive in Ogoni. Historically, gerontocracy did not silence the youth and women who both had their place in society. When compromised elites placed their selfish interests over the general good, the former elevated quiescence beyond its traditional boundary. Thus, in the face of environmentally disastrous events occasioned by Shell, KAGOTE kept quiet and no one else was expected to speak.

Ogoni pre-existing codes and traditions shaped MOSOP as the organisation began to draw on decades old tradition of organising, pre-existing cultural codes and structures. Thus, MOSOP must be seen as a reflection of the Ogoni cultural pattern. Similarly, Ogoni frames and strategies were shaped by broader national cultural codes such as democratic participation and federalism. A group of Ogoni scholars were persuaded that Ogoni ancient tradition, which was based on inclusiveness, and sustained Ogoni independence and prosperity provided the best pedestal for MOSOP. Prioritisation of inclusiveness constituted a challenge to the exclusive bent of KAGOTE and MOSOP. By adopting inclusiveness, Saro-Wiwa destabilised the rule of the elders and the ethos of quiescence in Ogoni. Thus, MOSOP is properly understood as a cultural challenge with direct continuity with national cultural values and Ogoni cultural ethos. As a result, the Ogoni conflict is better captured as a cultural challenge rather than a reaction to systemic dislocation.

8 Mobilisation

A place for moral motivation?

Introduction

In his highly polemic book, *High Stakes and Stakeholders*, Omeje typifies Nigeria as a 'rentier space', defined by 'high stake rentier politics', or a political tradition of desperate tendency to accumulation. The general disposition to 'political issues, discourses and phenomena are pervaded by high stake rentier mentality, calculations and manoeuvres'. This entrenched political discourse configures oil companies and communities, and finds abundant expression in civil society. The rentier space breeds and is characterised by violence and other destabilising tendencies (Omeje 2006: 6). This bleak categorisation of an entire nation raises at least two important questions. If the country is so compromised, where will the impetus for change come from? Elsewhere, the author argues that no society can be as irredeemably selfish and wanton as Omeje paints. His high-staked reductionism of the politics of elites and the masses to accumulation and survival struggles respectively is ahistorical and limits the possibilities of political action (Agbonifo 2009).

More importantly for our purpose here, in Omeje's rentier world, political action is motivated exclusively by an incorrigible penchant for primitive accumulation. Self-oriented calculations define the positioning of all societal actors, resulting in violent conflicts. The question arises; do moral motives have any place and role in the explanation of grassroots collective action in the Niger Delta? This chapter explores this question vis-à-vis the portrayal of the motivation of Ogoni activists in the literature and the latter's self-understanding.

Self-interest or moral motivation?

Saro-Wiwa argues for a federal system that assures equality for all ethnic groups regardless of size, and blames past leaders for betraying the true values of federalism. He emphasises that 'proper, undiluted federalism' only is what can save Nigeria from future destruction (Saro-Wiwa 1995: 76).

Okonta (2006) argues that the Ogoni became tribesmen in the attempt to attain citizenship. In a manner that emphasises an earlier suggestion of secession, Okonta claims Saro-Wiwa tended towards creating a nation-state or something near it. For such a bold claim he cites no credible authority other than referring to the emergence and proliferation of the Ogoni anthem, flags, calendars, handbills and colour posters with the photo of Saro-Wiwa.

Moral sense

What counts as a moral motivation in political action? Intuitionists explain moral knowledge in terms of reason, and locate immorality and selfish drives in passion. If reason tells us what is wrong, how does reason provide a motive to be moral? Does the answer lie in an external force that compels people to act morally? Hutcheson argues that human beings possess more than the five senses (cf Monroe et al. 2009). They have internal senses, including a sense of honour, beauty, the ridiculous and a 'public sense'. The latter ensures feelings of being pleased by the happiness of others and troubled over human misery (Monroe et al. 2009: 616). These internal senses are inborn just as the five senses. They cause us to react instinctively to approve good deed and disapprove bad acts. It is based on disinterested benevolence. Custom and education refine the moral sense. However, a natural substrate pre-exists them, otherwise it would be impossible to perceive moral right and wrong.

Thus, Monroe et al. (2009: 269) argue 'human beings are born with some innate needs and proclivities that encourage moral action'. While this may vary with culture and individuals, it has certain primitive components that cut across cultures. These 'components include, at a minimum, an innate need to feel benevolence toward others and to feel empathy for those in need or at risk' (Monroe et al. 2009: 269).And an innate need for self-esteem, which is tied to how we feel about how we treat other people. Therefore, self-interest is not the only or necessarily dominant human drive.

Monroe poses the question of what drives moral action? He argues that identity constrains moral choice. Identity refers to a sense of who I am, both as an acting and reflective individual. Explicit choice is deemphasised in favour of habits and a person's unique worldview. Such habits reinforce and form the individual's character. Good habits lead naturally, 'almost without thought' to do the right thing (Monroe 2001: 419).

It is pointless to develop moral character if that character cannot find expression in moral acts. Actors involved in moral activities are not different from others because they hold some peculiar moral beliefs, 'but rather by the high degree to which these values are integrated into the participant's sense of self. It is the integration that is key' (Monroe 2001: 421). The extent to which moral beliefs are embellished with how people see themselves is predictor of moral behaviour. The rescuers who took personal and mortal risks to protect Jews during the genocide had no choice but to help because 'doing anything else would have produced an inconsistency that would have rendered the self incoherent. What was at stake for the rescuers, then, was their very sense of self, their core identity' (Monroe 2001: 421).

The rescuers were imbued with such values as compassion, tolerance and respect for life. With these values well integrated into their sense of self, rescuers saw a connection, a human bond between themselves and the Jews, which others did not perceive.

> This altruistic identity and perspective created a sense of moral salience, a feeling that it was imperative for the rescuer to help. This feeling of moral salience meant rescuers could not turn away from Jews without turning away from themselves.
>
> (pp.421–422)

Teske (1997) deepens our understanding of moral motive. Moral behaviour is not entirely self-denial; to the contrary, moral behaviour stems from self-regarding concerns and concern for something outside the self. Teske posits that the rational approach shapes how some engage with political activists. Reactive to the rational actor school is a trend that emphasises the role of moral motives in political life. The latter impulse contains two strands: first, is the view that stresses the role of non-self-interest and altruistic motives in politics; and the second examines moral motives as a complex interaction between moral and self-regarding motives, which he refers to as an 'identity-construction' approach to moral motives.

The approach highlights how politics develop and express the identities of political actors and enables them to become something or someone that they otherwise would not have become. In the identity–construction approach moral motive is not marshalled as inherently opposed to self-interests as in the 'dual' motivation theory. Instead, it emphasises the ways that the construction of one's sense of self in politics is itself a moral project. Moral motivation and political behaviour is neither based exclusively on selfish costs/benefits calculations, nor solely on altruistic impulses. The identity–construction approach engages with concerns that are morally relevant and self-regarding at the same time (Teske 1997: 74).

Touraine (2000: 90) suggests that moral motivation is not limited to individual actors. Moral motivation can be found among certain social movements too. He argues that a social movement is never reducible to the defence of the interests of the dominated. Its ambition is always to abolish a relationship of domination, to bring about the triumph of a principle of equality, or to create a new society, which breaks with the old forms of production, management and hierarchy (ibid.: 92). Touraine asserts further that there is a prevalence of such movements, which challenge 'the modality of the social use of resources and cultural models'. The movements manifest a 'Subject' 'struggling against the triumph of the market and technologies, on the one hand, and communitarian authoritarian powers, on the other' (ibid.: 89).

The Subject is manifest by the presence of ethical values that come into conflict with the order of things. A societal movement defends 'a social modality of the use of ethical values, and comes into conflict with the modality that its adversary is trying to defend or enforce' (ibid.: 95).The ethical discourses imply freedom, and basic rights, which cannot be reduced to material gains. In a way that echoes Teske, Touraine emphasises the importance of not confusing the ethical dimension with the discourse of demand. The Subject is neither solely ethical nor material. Ethical discourses imply freedom, and basic rights, concepts irreducible to material gain. While collective actors may appeal to self-oriented needs, we need to be attentive to its moral appeal too. It is essentialist to reduce such an actor to its discourse of demand. Thus, Touraine's Subject resonates with Teske's identity–construction approach.

Environment: motivation and mobilisation

Earlier in the book, the argument was made about the intermingling of the Ogoni environment and community. Thus, the land is god and hosts many spirits that are venerated. The forests, beyond its trees and animals, are sacred, and as a result indiscriminate deforestation is unimaginable. Similarly, the human and non-human animals are coextensive. The essence of a man or woman could leave its own body and possess an animal (Saro-Wiwa 1992: 12). Such animals as the elephant, tiger, antelope, crocodile and catfish are designated as 'were' creatures. The Ogoni believe that when a 'were' is harmed; it has a ripple effect on the Ogoni people.

In other words, the environment is intricately bound up with the community. Thus, in their attachment to the land or abode of their ancestors, their sacred places and the roots of their existence, physically and emotionally, the Ogoni developed a sense of home. The Ogoni geographic location is more than mere space; it is a 'homeland', an historic territory, a heritage passed down by the ancestors through the generations, and, therefore, the rightful possession of the Ogoni people of today (Saro-Wiwa 1992, 1995). The Ogoni evolved a sense of attachment to their life-sustaining and spiritual heritage.

Ogoni land, forests and rivers are sacred; they evoke memories of bravery, the interventions of ancestors and deities, and a glorious past. All of these are commemorated in folktales, festivals, songs and masquerades. The Ogoni national identity is bound up with memory, which is rooted in the Ogoni environment. The land as the material embodiment of the Ogoni, therefore, lies at the root of Ogoni identity, 'community memory' of their past, present, and future, Ogoni prosperity, and guidelines for negotiating the world (Livesey 2001:73). It is against such deep attachment that one can understand Saro-Wiwa's moral outrage expressed in his dirge for the death of his beloved homeland: 'Where are the antelopes, the squirrels, the sacred tortoises, the snails, the lions and tigers which roamed this land?' (1992: 83).

The ghosts, real and imputed, that constitute the distinctiveness of the Ogoni environment, and fostered a sense of attachment or elicited emotions of beauty and joy, were under assault. Processes of development were divesting Ogoni of its salient place characters, with grave implication for the Ogoni identity. Appalled by the destructive activities of Shell and insensitivity of the State, the Ogoni in 1990 deliberated on their condition and, in revulsion, decided to do something about it. The decision was motivated by the Ogoni sense of who they are in relation to their environment and who they are (equal to any other) in relation to other ethnic groups in Nigeria. To walk away from the destruction of their environment by capital was to walk away from their ancestors, deities, guidelines for negotiating the world, and, indeed, from themselves. Similarly, to acquiesce to continuing political marginalisation and exploitation was to accept and imbibe imposed categorisation as 'marginal' and 'less-than-equal'; the very opposite of the identity of equality.

These identities are not new. The Ogoni have a long and rich history of struggle for self determination in Nigeria. In that struggle, T.N. Paul Birabi, foremost Ogoni nationalist figure, serves as a role model. Birabi has been used as a historical figure by activists. Birabi remains salient in the memory of the Ogoni, and he is regarded as a reputable forebear who undertook heroic service for the Ogoni nation. Ogoni activists assert continuity with Birabi's dogged political struggle for Ogoni self-determination, and contrast the figure's selfless service with the selfish tendency of a few. This revered figure, and others such as S.F. Wika, the Reverends Wiko and Badey and Bishop Vincent, serve as Saro-Wiwa's role model (Saro-Wiwa 1995). Saro-Wiwa states emphatically that he has endeavoured to emulate Birabi in his entire struggle for the Ogoni.

Habitual behaviour encouraged by significant role models enhances habits of caring that can be moulded into altruistic personality (Monroe 2001). In the case of the Ogoni, these habits included fighting perceived injustice and oppression. 'My worry about the Ogoni has been an article of faith' (Saro-Wiwa 1995: 49). Saro-Wiwa actualised his concern and care during the Nigerian civil war in 1967–1970, and as a member of the Rivers state executive Council in 1968–1973. During the war, he took action geared at rehabilitating the displaced Ogoni people. As Commissioner in Rivers state, Saro-Wiwa awarded scholarships to qualified Ogoni students. In a pamphlet released in 1968, during the war, Saro-Wiwa expressed concern to unite the Ogoni and ensure that Ogoni 'regains its lost dignity and honour, and transform our land for the betterment of our peoples' (pp.52–53). Underlining the influence of Birabi on his character and moral behaviour, he asserts 'The spirit of self-sacrifice which moved Birabi is still alive in our nationality today' (p.53).

Saro-Wiwa, a student of society, also drew inspiration from reputable public figures from other parts of Nigeria. Copiously, he quoted from Nigerian nationalist leaders, Chief Obafemi Awolowo, a Yoruba, authoritative pronouncement on ethnic equality and true federalism. On collective mobilisation, Saro-Wiwa claims he learned a good lesson from historical mobilisations in Nigeria. Awolowo, Nnamdi Azikiwe (Igbo) and Ahmadu Bello (Hausa-Fulani) successfully mobilised their people on ethnic platform; 'I thought I could do the same for the Ogoni' (Saro-Wiwa 1995:101).

A writer must be engaged politically in reaction to his/her situation. A writer in a society such as Ogoni must engage in activism; it would be 'irresponsible' to do otherwise. A writer 'must have the courage of his conviction and be prepared to take responsibility for his actions and believe with the ancient saying that, Magna est veritas et praevalebit – for great is truth and the truth shall prevail' (ibid.). To foreground his conviction, Saro-Wiwa claims that the best Nigerian writers have never shied away from such engagement. He refers to Nobel Laureate, Wole Soyinka, Chinua Achebe, Chris Okigbo who died fighting on the side of Biafra, and Nigerian academic Festus Iyayi who was actively involved in organised unionism and the democracy movement. Thus, he discounts the value of those who only write and watch; 'I am only reacting to my social situation, as every writer of my value must' (1995: 82). Given such values, Saro-Wiwa naturally reacted to the 'Injustice', which 'stalks the land like a tiger on the prowl' (1995: 8).

The question might be asked how the passion or moral motivation of Saro-Wiwa can be extrapolated to the Ogoni who mobilised under the banner of MOSOP. Ogoni historian and academic, Kpone-Tonwe claims that the Ogoni cultural club, KAGOTE, was aware and concerned about the plight of the Ogoni, which explains why the group accepted the idea of the OBR and MOSOP. However, conservative members of KAGOTE did not completely agree with the strategies of Saro-Wiwa. Moreover, some have argued that there is no pan-Ogoni identity. However, Michael Zuckerman (1978: 339) argues that there are times in history when 'some men transcend their situations, and times more striking still when substantial numbers converge to some shared transcendence'. How did substantial numbers of the Ogoni come to imbibe an insurgent identity with a moral dimension? How do we bridge the seeming gap between the moral temperaments of one activist leader with the entire movement?

To galvanise mobilisation, political movements against domination must develop new diagnoses that raise suffering as immoral and unjust, and suggest a remedy for extant forms of suffering (Moore 1978). A significant number of actors need to agree on a collective definition of a condition as in need of ameliorative action and persuade others to accept that definition of reality as true. The process of constructing reality involves the use of framing activities and generation of vocabularies of movement motives. Frames refer to 'the conscious strategic efforts by groups of people to fashion shared understandings of their world and of themselves that legitimate and motivate collective action' (McAdam et al. 1996: 6). Collective action frames, organises experience and guides action in ways intended to activate adherents and convert bystanders into followers (Snow 2004). 'Frames' make a strong case for the 'injustice' of a given condition, who to blame, what should be done to change the situation and why collective effort would be successful.

A sense of moral duty is essential to action mobilisation. Movements are involved in the motivational task of constructing and amplifying beliefs about why it is proper and moral to take action to ameliorate the vexing situation. Critical to that effort is the elaboration of the identity of protestors. Such definition precedes the identification of interest. Interest definition requires a solid basis of self identification to rest on. The Ogoni elaborated an oppositional identity to the State and Shell, who were defined as exploitative, unjust and wicked. For activists, the Ogoni is Ogoni and Ogoni is the Ogoni; the Ogoni have equal stakes in the Nigerian federation as any other ethnic group, big or small, and they have a right to protect their God-given land. To solidify the identity, activists appealed to the ancient valour, historical figures, independence, self-sufficiency, prosperity and happiness of Ogoni prior to colonisation. In articulating their identity, the Ogoni appealed to moral and nationalist values that were more than provincial.

From a social psychological perspective, identity exerts a direct and indirect effect on collective action participation. Directly, people who identify with a group or cause may participate in a struggle not necessarily for an instrumental reason. Indirectly, collective identity can mollify instrumental or strategic logic such as to make a 'free ride' less attractive. In other words, 'High levels of group identification increase the costs of defection and the benefits of cooperation. Collective identity has an impact on the instrumental pathway to protest participation' (Klandermas et al. 2002: 236). The argument provides a certain resolution of Olson's (1965) free-rider problem: why rational actors would engage in collective action when they could free-ride. However, the authors sustain the dualistic thinking inherent in the argument and, which is taken further by the view that participation generated by identity considerations result in automatic behaviour, whereas participation informed by instrumental reasoning is deliberative.

Binary thinking reduces variety of reality to just two poles, constraining the possibility of asking important questions about what lies in the interstices of the polar extremes. If action is informed by either instrumental or identity logic, how may we categorise behaviour generated by both identity and rational calculation? This is an important question if we accept that there are no pure motivational types, but only hybrids. In the following, an attempt is made to show how analytical categories of self and moral motivations fail to measure up to reality.

Activists' self-understanding

Empirical analysis of interviewees' articulation of why they joined the movement and activists' texts points to three important lessons. First, the fluid weaving of self-interest and moral motivations in a way that defies attempts to categorise activists' motivation either as selfish or altruistic. Second, a sense of who they are, or identity, shaped activists participation, and reaction to military repression. Third, activists' identity did not emerge preformed, but was developed in the course of collective action.

Fluid weaving of self and moral motivations

In the OBR, the Ogoni, without equivocation, articulated the country's federal system as skewed, and therefore, the problem with Nigeria. They accuse successive governments of having betrayed the true values of the State through trampling on minorities' Constitutional rights, and use of naked power to transfer Ogoni wealth to other parts of the Republic. Thus, MOSOP positioned itself within the larger articulation of dissatisfaction with the nature of Nigeria's federalism, including the power relations between the centre and minority groups and between the minorities and majority ethnic groups (MOSOP 2004: 5–6). Employing State power to deny the Ogoni and other minorities their rights makes the 'beloved country a very unequal one or, for the Ogoni andtheir like, a slave society in which the master groups have all and the slave groups nothing' (MOSOP 2004: 45–46).

At the same time, the OBR deploys a discourse of demand, bemoaning the lack of basic social infrastructure, absence of schools, hospitals and piped water. In consequence, the Ogoni demand compensation for environmental degradation, payment of royalties, and concessions giving the Ogoni access to oil revenue for their own development, political autonomy within Nigeria, and right to protect Ogoni environment against further degradation.

> We make the above demand in the knowledge that it does not deny any other ethnic group in the Nigerian Federation of their rights and that it can only conduce to peace, justice and fair play and hence stability and progress in the Nigerian nation. We make the demand in the belief that, as Obafemi Awolowo (Nigerian nationalist) has written: In a true federation, each ethnic group no matter how small, is entitled to the same treatment as any other ethnic group, no matter how large.
>
> (MOSOP 1990)

Saro-Wiwa remarks that MOSOP was formed in the attempt to extricate Ogoni from internal colonialism and environmental strangulation, and to challenge the exploitative and oppressive system imposed by the military on Nigerians. He employs powerful concepts to flesh out the deep roots of the Ogoni struggle. The concepts are powerful because they do not adhere well to the narrow confines represented by redistribution and recognition. For instance, there is nothing about redistribution or recognition in terms such as 'cruel', 'insensitive', 'primitive', 'indigenous colonialism', 'democracy' or 'progress'.

> The call for self-determination was therefore a ... rejection of indigenous colonialism which I have characterized elsewhere as cruel in the extreme, insensitive and primitive. It must stand rejected in the interest of social progress, for it is this colonialism that is responsible for the backwardness of Black Africa. All over the continent are despairing, distressed peoples,

held in thrall by their kind who usurp their rights and subject them to the status of third-class citizens or outright slaves, thus destroying their culture.

(MOSOP 2004: 47–48)

We find in the discourse of the OBR a commingling of provincial and moral concerns. The OBR references self-regarding concerns by stressing the material lacks of the Ogoni and identifying their material demands. In the same breathe and without any sense of contradiction, the OBR demonstrates a moral concern when it appeals to Nigerian independent leader, Chief Awolowo, to argue for ethnic equality regardless of size as sine qua non for a better Nigeria. 'The present structure reinforces indigenous colonialism – a crude, harsh, unscientific and illogical system' (cf Saro-Wiwa 1995: 63). Expanding its moral tone, the OBR makes clear that Ogoni demand in no way detracts from the interests of other ethnic groups. To the contrary, it makes for the wellbeing of the federation. Moreover, it accused the government of betraying the true values of the State, which the Ogoni are intent on protecting. Similarly, the OBR couched its disquiet with Ogoni material destitution, despite being resource-endowed, in a moral grammar when it asserts that such destitution is 'intolerable'. Intolerable invokes a sentiment of disapproval, pain and unendurable event.

Similar sentiment of revulsion characterise Saro-Wiwa writings. In his book, *A Month and a Day*, Saro-Wiwa described the exploitation of the Niger Delta as a result of its relatively small population size as 'unjust, immoral, unnatural and ungodly' (1995: 64). He called on the Babangida administration to extend its human rights and social justice claims to minorities throughout Nigeria. Also, he appealed to national elites 'to show compassion to the less privileged of our society so that we can achieve a better Nigeria' (ibid.). But Saro-Wiwa's moral outrage did not end there; he explained that in an unjust situation such as Nigeria's the writer must reach out, participate in mass movements. In such a dire situation, 'it is idle merely to sit by and watch or record goons and bumpkins run the nation aground and dehumanize the people' (1995: 82). For Saro-Wiwa, it is idle not to be involved; with such a perception doing nothing would have been inconsistent with his sense of self, producing self-incoherence. Here, Saro-Wiwa makes explicitly clear Monroe's argument that what sets rescuers apart is the high degree of integration of values into a sense of self.

When Babangida created new states and left out the agitations of the Ogoni and others in Rivers state, Saro-Wiwa asserts

it was this act alone which so outraged me that I decided that, come life or death, the brutalization of the peoples in the oil-bearing delta of the Niger would have to be questioned, exposed and brought to a stop.

(pp.99–100)

Babangida's modus operandi was 'an affront to truth and civility, a slap in the face of modern history; it was robbery with violence...The brazen injustice of it hurt my sensibilities beyond description' (p.100). More than that, Saro-Wiwa opted to do something about the situation; he could not sit idly 'watching Nigeria literally go down the drain through the incompetence and banditry' of the rulers (p.100). Saro-Wiwa focuses attention on state-engineered material conditions of the Ogoni but deploys moral grammar to problematise the state's actions and the outcomes. Thus, the material basis of Saro-Wiwa's activism and the Ogoni demands become conceptually insepar-able from moral motives. The failure of the government to entertain Ogoni demand for a state relates to the Ogoni self interest. However, construing that failure as 'robbery' and 'violence' is decidedly a moral issue. Matching the Ogoni deprivation intimately with a vision of a country going down the drain, and about which something has to be done is simultaneously self-oriented and moral.

Ogoni academic Ben Naanen argues that joining MOSOP came naturally to him,

> Involvement in MOSOP was natural to me. I did not need any incentive and none was given. An issue of concern to me then as General Secretary of MOSOP, while undertaking post-doctoral studies at SOAS, concerned ethnicity. As I had already developed a pan-Nigerian view, I was con-cerned about narrow ethnic agenda. I later thought that both are not mutually exclusive – and that in the pursuit of national vision one needs a platform. With a single ethnic movement one could pursue the national vision, that is, a metaphor of ethnic agenda to pursue a broader national vision. That was how I resolved the dilemma. The impetus that galvanized all Ogoni leaders may not have been the same. However, the failure of conventional politics to salvage minorities was of general concern. And mass movement became imperative. MOSOP had a national view but with a core Niger Delta and Ogoni agenda, employing ethnic agenda, which was available. There was no other platform to ensure mobilization. We are motivated by altruistic ideas: we did not think about personal gains.
>
> The struggle has revealed to me the extent of injustice in this country, how difficult it is to deal with the state and multinational corporations. It has also exposed the influence Shell wields over the government. I realized both the state and Shell were totally unprepared for the ques-tions raised by MOSOP. For example, Shell had little awareness of cor-porate social responsibility; thus its policy change was a direct result of Ogoni struggle.
>
> (Ben Naanen 2008 int.)

To the scholar, ethnic and broader national orientations are not mutually exclusive. He believes that one can foster a national agenda on the pedestal of an ethnic agenda, especially given that such was the only available platform. Thus, it was clear to him that while MOSOP was Ogoni-oriented, it had a national orientation. His sombre appraisal of the State as insensitive to injustice, and privatisation of the State by Shell echoes concern expressed about the State and Shell in some scholarly quarters (Ake). Such concern is moral at the same time that it was self-interested.

Collective action and otherworldly

Activist vocabularies too, comprise moral motives and self-interest in such a way that one cannot begin to imagine untangling them. Saro-Wiwa attributed his activism to 'the Voice' of the Ogoni spirit with a command to liberate the Ogoni and others under oppression in Nigeria (MOSOP 2004: 44). In effect, participation in collective mobilisation was at the instance of divine will; not the private interest of the activists. Yet, the 'call' is an invitation to struggle for the rights of the Ogoni and non-Ogoni. Even within the call we find a mingling of self-directed and moral motivations. Similarly, many activists interviewed appeal to a conviction or sense of divine inspiration to undertake certain tasks for the movement. These activists express firm commitment to the struggle and bravery regardless of the odds. They wish that their lives would be meaningful in some way. Tanee explains that he was elected as leader of the Uegwuere Chapter. He accepted the election result because 'I thought maybe God wanted to use me' (personal interview 2008). Tanee aspires to die as an MOSOP activist or martyr saying that if he dies as an ordinary person, he would not rest well in the grave. While such is a concern with the self-being focused on his life, it remains pre-eminently a moral concern to be meaningful to something outside the self. It expresses a concern to be significant and useful to MOSOP, to die in the cause of the struggle, executing the divine cause.

Des Laka explains that his own activism has a 'mystical ring to it'. He claims,

> Saro-Wiwa appeared to me in a dream, and wondered why I had not returned to Ogoni to join the struggle. In 1995, I returned home and a woman took me to a meeting of MOSOP clandestinely held in the forest. Whenever I feel tired, something seems to push me on. The benefits I have derived have to do with the fact that I have a name known all over Ogoni. I am known for my transparency and honesty. I have joy doing what I do; the joy is more than material benefit.
>
> (Des Laka 2008 int.)

Charity Deekae Dickson claims,

> For now I don't gain anything from being a member of MOSOP. But I have retained membership because we are fighting a just cause. Any day we get our freedom, I will benefit. I was brutalized along with some other Ogoni students at a police check-point. My shoulder got dislocated and we were denied treatment. But I have no regret; God has a reason for

everything. My father was an activist. The day Saro-Wiwa was killed no one ate in our house. The Nigerian government is wicked and I am angry. Whatever sacrifice I can make towards the struggle I will do it. Only yesterday, I received notice of this meeting and I decided to suspend everything else. Even though I am grieving my brother who died few days ago and had to close my business shop to attend this meeting, I reasoned it was worth attending.

Charity Deekae Dickson is self-employed businesswoman. She stresses that the struggle has not brought any benefit but pain and personal sacrifices. Yet, she maintains her identity as a member of MOSOP because she believes it is fighting a just cause. Brutalized by the police at a checkpoint, she remains unbowed and continues to make sacrifices for the struggle in the hope that when freedom comes, she would benefit. However, she locates her commitment in the belief that God has a reason for everything that happens to her and to the Ogoni. Moreover, despite her bereavement, she made time to attend the meeting at which I encountered her because she reasoned it was worth it. Here, we find an appeal to personal interest in the event of freedom. That appeal is, however, mingled with faith in the omniscience of God and a firm belief that commitment to a just cause was worth it, in other words, moral and right.

Collective mobilisation and identity formation

MOSOP (2004) highlights its intent to transform the unorganised and quiescent Ogoni community into a 'mass-based discipline' movement that would draw upon its culture to re-establish the identity of 'self-reliant' community, shifting from an apolitical base to a challenger of tyranny. Activists deployed resonant frames with the intent to galvanise Nigerian communities to question the State and demand fundamental change in the order of things.

> MOSOP was intent on breaking new ground in the struggle for democracy and political, economic, social and environmental rights in Africa. We believe that mass-based disciplined organizations can successfully re-vitalise moribund societies and that relying upon their ancient values, mores and cultures, such societies can successfully re-established themselves as self-reliant communities and at the same time successfully and peacefully challenge tyrannical governments. MOSOP also believes that debt-ridden, morally-bankrupt Nigeria must be a federation of equal ethnic groups, irrespective of size, with each group being free to control its resources and environment and exercise its political right to rule itself according to its genius.
>
> (MOSOP 2004: 5–6)

The Ogoni demand was therefore meant to awaken the government of Nigeria to its responsibilities to the oppressed ethnic groups throughout the length and breadth of the country. The Ogoni chiefs and leaders were fully aware that there are many minority ethnic groups in all parts of the country who were suffering disabilities similar to theirs and hoped that whatever was done for them would also apply to their fellow-sufferers. In that sense, the Ogoni Bill of Rights was a broad-minded document, touching the fundamental rights of all Nigerian communities and meant to initiate fundamental change.

(MOSOP 2004: 51)

MOSOP gave rise to the Subject by emphasising ethnic equality, freedom of ethnic groups to control their environment and 'fundamental rights of all Nigerian communities'. In the same vein, MOSOP's discourse has a tinge of self-interest expressed in the Ogoni 'hoped that whatever was done for them'. That whatever included self-regarding material demands (OBR). Yet, it is grossly essential to define MOSOP by its discourse of demand alone. What provoked the emergence of MOSOP was moral outrage over environmental strangulation and political marginalisation. The Ogoni resolution strategy was to question the order of things, not merely demand reforms, and press for fundamental change. The movement sought to abolish a relationship of domination at the same time it sought provincial benefits.

In the attempt to make explicit what the MOSOP was for and against and thereby galvanise grassroots mobilisation, the miideekor frame was suggested. The frame resonated with the entire Ogoni very quickly and became one of the most powerful tools for expressing Ogoni identity and demands. In the Ogoni worldview, the tenant, or palm wine tapper, who fails to deliver the landlord's miideekor, is a 'thief', and the landlord is entitled to reclaim his palmfield. On face value, the elements of the symbolism are material or economic. The symbolism itself, however, signifies something deeper: failure for any reason to pay miideekor undermines social expectation and amounts to denying the landlord his rights. Given this double constitution of miideekor, collective action aimed at wresting control of the field from the palm wine tapper or extracting the miideekor is simultaneously economic and cultural, self and moral motivated. Thus, in the course of collective mobilisation, the Ogoni developed a new identity as landlord with rights, and took action to enforce those rights.

Individual activists emphasised how they developed character in the course of mobilisation. For instance, Tanee argues:

I enjoy the life of an activist. I am more experienced now. I have learnt a lot of lessons, which I am able to teach to others. The struggle has changed me because now I know the difference between good and bad. In the struggle, I am no longer afraid of people provided I am in the right. I am against corruption, bribery and deception of people. The struggle is all about the future of our people. If Ogoni benefit from the struggle, then I will benefit. If I am no longer here, my children will benefit.

(20 March 2008 interview)

Tanee puts his finger on decidedly private or personal benefits. However, notice the link he draws: following his experiences, Tanee is able to impact knowledge to others. Given his new capacity to apprehend good and evil, he has become courageous. He has become a certain kind of person as a result of the struggle. Both virtues translate to shape his interactions with others. A sense of morality, or dispositional response to others, not strategic calculations, shapes his social interactions.

One learns from activists what Teske terms the incomplete dichotomy between moral and self-motives. Self-related and moral inclinations mesh easily forming a seamless tapestry in activist's discourses. Tanee is self-confident; he enjoys activism, abhors corruption and is more politically experienced because of activism. These personal benefits enable him to lead boldly and legitimately. They provide him materials with which he teaches his followers, and he is able to lead transparently and forthrightly. While character development conveys a notion of self concern, it also has to do with something external to self; he is able to provide effective leadership to others.

The struggle changed Tanee; he is against corruption and deception, underscoring how activism has been a process of character development. Given his new lifestyle, he is able to walk freely everywhere without fear of molestation or accusation of graft. These are some of his rewards. Such rewards also have moral content. For instance, because of his transparency, he is able to undertake micromobilisation all over Ogoni. Moreover, people see him as honest and as a result listen to him, making him a better leader.

Nwigani is a secondary school teacher, and women leader in MOSOP. In the course of the struggle, Nwigani has travelled overseas and participated in different political meetings and conferences exchanging ideas. As a result, she has become more politically conscious and knowledgeable. Nwigani emphasises:

> I have benefited from the struggle. Initially, I did not know Shell was cheating us but now I do. As a result of the struggle I have been privileged to attend seminars and conferences overseas. I am a teacher by profession but the struggle has made me a teacher with a difference because I can now use examples and experiences from the struggle to instruct my students. I am also now very self-confident as an Ogoni, and other people now come to us for advice.
>
> (2008 interview)

Such satisfaction must relate to moral motivation because it shapes her public function as a schoolteacher. Her students are arguably better off because of her new experiences and psychological satisfaction. Thus, Teske argues that the attempt to squeeze political motives into a category labelled 'self-interest' or 'moral' will end up in frustration. Such effort is ill advised because much of the satisfaction or the rewards reported by interviewees and others are entangled in a way that defies the dualism of self/other.

Conclusion

Outrage over environmental despoliation derived from the Ogoni sense of attachment to the environment. Environmental pollution was seen as constituting a destructive and divisive process to the interconnection between Ogoni deities, the land and the Ogoni. However, it was also partly the result of distribution issues vis-à-vis other ethnic groups in Nigeria.

Yet, even that imagination is place-shaped because the spatial marginalisation of Ogoni further concretised the sense of how actors from afar and near shape the fortune of Ogoni. That development was an affront to the history of Ogoni valour and autonomy. The materialisation of exploitation in Ogoni was seen by the Ogoni as ahistorical, and out of place. The resulting outrage was an impetus to fight and re-establish the status quo ante.

9 Place and limit of mobilisation

Introduction

The Ogoni are spread across six traditional kingdoms, all nestled together within an area measuring only 404 square miles (1,050 km^2). The open topography impressed on MOSOP leaders the futility of armed struggle when the movement received offers of ammunition from external sympathisers to the Ogoni cause (Ben Naanen 2006 int.). The leader of the Rivers State Internal Security Taskforce, whose role was the pacification of the Ogoni, Major Paul Okuntimo, attests to how the geography of Ogoni facilitated his troop's effective assault strategy against the Ogoni:

> Nobody knew where I was coming from. What I will just do is that I will just take some detachments of soldiers, they will just stay at four corners of the town. They ... have automatic rifle[s] that sound death. If you hear the sound you will freeze. And then I will equally now choose about twenty [soldiers] and give them ... grenades – explosives – very hard one[s]. So we shall surround the town at night ... The machine gun with five hundred rounds will open up. When four or five like that open up and then we are throwing grenades and they are making 'eekpuwaa!' what do you think ... the people are going to do? And we have already put roadblock[s] on the main road, we don't want anybody to start running ... so the option we made was that we should drive all these boys, all these people into the bush with nothing except the pant[s] and the wrapper they are using that night.
> (HRW 1995)

Accessibility conferred on the soldiers the advantage of entering Ogoni from different points. There is a hint that the Ogoni terrain was open and accessible in Okuntimo's stress on roads, road-blocks and people running into the bush, which contrasts sharply with the watery, mangrove and swampy terrain of the Ijaw militants. According to HRW, raids on Ogoni towns and villages followed a pattern; troops invaded a community shooting indiscriminately. As frightened men, women and children fled to the surrounding bush they were shot at by the invading force. Then, the soldiers and mobile police attacked houses, breaking down doors and windows. Those they met in the process were severely beaten, raped, shot and/or compelled to part with money.

Terrain and militant groups

In November and December 2005, Ijaw militants detonated two explosives in the creeks of Rivers State, destroying two Shell Petroleum Development Company (SPDC) pipelines located in Okirika and Andoni (Coulson 2009: 18). On 11 January 2006 an SPDC oil field located about 20 km offshore was attacked, damaged and four expatriates on site were abducted after a fierce gun duel between militants and military guards at the facility (ibid.). On 15 January 2006, militants 'attacked and destroyed one flow station and two military house-boats belonging to SPDC in Benisede, Bayelsa State' (Coulson 2009: 18). The attacks and many thereafter were masterminded by the Movement for Emancipation of the Niger Delta (MEND). The objective 'is to totally destroy the capacity of the Nigerian government to export oil' (ibid.). These attacks and shut-in of 400,000 bpd led to a hike in the price of oil on the international market (Coulson 2009).

In official quarters, militant groups in the region, including MEND are criminals bent on undermining the national economy through illegal oil theft and other illegal activities. A leaked military report by the then commander of the Joint Task Force in the Niger Delta, former Brig. Gen. L.P. Ngubane, listed groups such as MOSOP, INC, IYC, NDPVF and FNDIC as militant groups that must be crushed. Identified for total elimination are Ijaw communities such as Oporoza, Kurutie, Kunukunuma, Okerenkoko and other villages in Gbaramatu clan, Delta state, seen as hotbeds of militancy (Coulson 2009). It is no surprise that there has been an increase in the use of the state security forces to crush MEND.

Beginning on 15 February 2006, military helicopter gunships bombarded Ijaw communities for three days, leading to the death of 20 persons and the destruction of three communities, namely Perezuoweikorigbene, Ukpogbene and Seitorububor, in Gbaramatu clan. MEND interpreted the aerial assault as a premeditated targeting of the communities seen as sympathetic to its cause. The group mounted a reprisal attack against the Forcados oil export terminal and took nine expatriates hostages. According to Jomo Gbomo,

> The location was chosen as a target to show the oil companies, military and the government that they (MEND) have the capacity and 'facilities' to storm any oil facility in the creeks of the Niger Delta and render it impotent not minding how many military personnel are deployed for its protection. The group also promised to 'continue to nibble the Nigeria oil export industry until we think necessary to deal it a final blow. We have caused the oil companies and Nigeria government to pay more for our oil and eventually, it will be snatched right out of their grip'.

(cf Coulson 2009: 21)

Again, in May 2009, the military deployed four jet fighters, 24 gun boats and three battalions of the Nigerian army into the area (Coulson 2009). According to Coulson, the JTF attributed the renewed hostility to the killing of 18 military officers guarding oil facilities by fighters loyal to a militia leader Government Ekpemupolo (Tompolo). Camp 5, a suspected stronghold of the Tompolo-led militia, was attacked by the military using jet fighter and bombs and aided by ground troops and naval support. The bomb attack was extended to Oporoza, the headquarters of the Gbaramatu Kingdom where a cultural festival attended by indigenes and visitors was ongoing. The military cordoned off the waterways and creeks then raided suspected militants' camps in the territory. The military air, land and sea attack on communities left several innocent people dead. In retaliation, the militants targeted oil facilities in the territory by blowing up pipelines, flow-stations and oil facilities with the intent of crumbling the oil economy (Coulson 2009). The attacks reduced the oil output from 2.6 million barrels per day (bpd) to 1.8 million bpd within a month of renewed militia attacks on oil facilities (ibid.).

> There are a multitude of reasons why the Nigerian government has been unable to stabilize the Niger Delta. One of the most obvious explanations is the terrain of the delta. According to the Niger Delta Development Commission, the delta is the world's third largest wetland and is composed of dense mangrove swamps and waterways, making it an ideal location for guerrilla operations. The various oil facilities and pipelines saturate the area and are easy targets for militants who are able to navigate the dense web of waterways in speedboats, lay siege to a facility, capture international oil workers and then disappear back into the swamps and mangroves. The speed and size of the guerrilla attacks often catch the security forces protecting the energy installations by surprise; these same security forces usually suffer from poor equipment, training and morale, placing their dedication in doubt. The weapons used by the militants are abundant in the country since small-arms filter into Nigeria from conflict zones like Liberia, Sudan, Somalia, the Democratic Republic of Congo and Sierra Leone.
>
> (Marquardt 2007)

The capacity to hide from security forces and accessibility provided by the maze of creeks may have informed why militants decided to relocate to the mangrove jungle. Coulson (2009) buttresses the argument when he asserts that MEND militants relocated from the cities into the creeks, realising that the disruption of oil exploitation activities would have domestic and global impacts. According to Coulson, MEND does not have a single command structure. Instead, it is an amorphous organisation with dispersed leaderships across the Niger Delta. Jomo Gbomo, a pseudonym, was the group's only known face who was only accessible through the mass media. Such an organisational model was germane to effective guerrilla warfare extending over the whole region. The rationale for the model was to avoid top leadership being targeted for elimination as happened in the case of MOSOP. The 'invisible' nature of MEND is an important factor making it difficult for the government, oil companies and even the military to target the organisation and effectively neutralise its activities in the Niger Delta.

Osaghae remarks on the effectiveness of the disruptive actions of MEND, arguing that the destabilising consequences of the regime of disruption, political and economic instability in the region orchestrated by the militant movement, compelled the federal government to embrace reconciliatory measures (Osaghae 2008)..

In other words, the less destabilising strategy of the Ogoni was insufficient to elicit the State reconciliatory gesture. But this begs the question how a relatively poorly armed group could undermine the awesome might of the Nigerian army in an asymmetrical conflict that favoured the latter (Tschirgi 1999). What factor facilitated the unhindered disruptive activities of the militants and prevented the state security apparatus from arresting or destroying them? This is an important question because the military was on the trail of the militants in the Niger Delta. So, if the military had the opportunity they would have eliminated the militants and destroyed their organisation as it did to Ogoni leaders and MOSOP. Like many concerned scholars, Osaghae fails to draw the link between difficult terrain and the militants' audacious disruptive practices.

Terrain, which provided 'safe spaces' for militant groups to inflict serial damage on oil infrastructure and by extension, Nigeria's economy, and to hide from reprisal attacks is a signal difference between militant movements and MOSOP. At will, the State could easily reach out, grab Ogoni leaders and subject the same to violent sanctions; it could not do so in the case of militants despite the massive deployment of ground, water and air troops. This is taken as the major determinant and big gap between the movements because after the demise of Ogoni leaders and totalistic repression, the movement lost its Saro-Wiwa inspired vitality (Nepstad and Bob 2006). Military action led to the death of some militants and their commanders (AA, 2011 int). However, the core leaders of the militants remained invisible and invincible. Thus, they continued to provide the leadership and inspiration the rank and file needed to sustain the violent struggle.

While disruptive tactics could be costly and simultaneously successful, in the case at hand, the militants maximised the positive benefits of disruptive tactics and minimised the costs, but the relatively less disruptive action of the Ogoni reaped the maximum penalty of disruptive action. What explains the critical difference between both movements lies in the place where disruptive activities happen. Eight informants belonging to different militant organisation were interviewed and they all emphasised that the difficult mangrove terrain of the Niger Delta provided hiding places and advantageous positions to initiate violent attacks on oil facilities and military personnel. Samples of militants' articulation of the critical role of geographic terrain to their capacity to escape repression, arrest and violent death, presented below, buttress the argument (respondents are anonymised).

> We understand the rivers and the creeks very well. We work, fish and live there. Everywhere is bushy, so many creeks. So, when the soldiers are coming from the right, we divert to the left. We got information on the soldiers' movement from our people on the ground. We also have among us some old (retired) soldiers who used their military and creeks knowledge to help us hide from security forces.
>
> (JC, 2011 int.)

> Some creeks are very difficult to identify and enter; even canoes cannot enter them let alone boats. You must enter the creeks before you can access the camps, which are located deep inside the difficult creeks. Because they are concealed, the army finds it difficult to determine the location of the creeks. The nature of the creeks confuses the soldiers. But we know it. This helped us to move and carry out our operations successfully. But fetish traditional beliefs also provided us protection. Sometimes, rivers overflow and cover the land. So, it is difficult for the army to locate the exact point of access to the creeks or where the creek is. But we know because we grew up there. It is our area. We are fishermen. Those camps belong to our fore-fathers. They used them for fishing. Outsiders cannot find them. Even aerial view from helicopters cannot locate some of the creeks. It is difficult for carnal (natural) man to see or locate them because such areas are spiritually and specially designed.
>
> (ME, 2011 int.)

> Being in the creeks gave us advantages over the military. The creeks created safe havens for us because there was no government there; we were the government, we were like Robin Hood. Creeks are inhospitable places; only God saves those who live there. So, it is difficult to send a policeman there to arrest somebody. Blowing up pipelines was our specialty. The mangrove swamp is a jungle. Where we stayed is called a camp. The JTF cannot navigate the creeks. Some of the creeks were dug up by us. Some of the creeks and canals are not on the map because we created them. We are under the vegetation, so, you cannot see us from the air. This natural vegetation helped us a lot. We could set traps to torpedo their gun-boats. The army was better armed than us, but in terms of guerrilla tactics we were superior to them. And the mangrove swamps helped us a lot.
>
> (EK, 2011 int.)

The militants occupy the difficult but vast watery, swampy mangrove vegetation of the Niger Delta. Their knowledge of the difficult terrain was exploited with devastating effectiveness against government troops. This vast and difficult terrain provided safe spaces in which they could plan and mount surprise attacks against oil facilities and armed troops. Importantly, the terrain enabled them to disappear and hide following such audacious attacks and or when better armed units of the troops were on their trail. With perfect intelligence on their side, the militants developed the ability to attack oil facilities and troops at times and places of their own choosing. Surprise and ambush gave them certain advantage over the troops. The basis of such advantage was that the difficult terrain was the insurgents' home; they knew the nooks and crannies of the topography like the back of their hands.

Conclusion

The outcome of social movements in Non-Western countries has been given little notice in the literature. In particular, there is scant attention to the outcome of protest activities in the Niger Delta. The chapter seeks to fill in the theoretical gap. The argument develops through systematic analysis of the trajectories and outcomes of two protest movements in the Niger Delta; the Movement for the Survival of Ogoni People and militant movements. Both movements emerged in response to broadly similar conditions: political marginalisation and environmental despoliation. While MOSOP preferred non-violent disruptive grassroots mobilisation; the militant movements ab initio adopted armed insurrection. Both movements utilised, albeit, to varying extent, disruptive actions, including protest marches, violent actions, and confrontation with Shell and armed troops. Despite the differences between them, the reaction of the Nigerian state was largely the same. The State's violent reaction suggests it was less the character of the movement than the nature of the State that determined movements' repression.

Military repression of the non-violent disruptive struggle of MOSOP led to the judicial murder of Ken Saro-Wiwa and eight other Ogoni leaders, long-term incarceration of many, summary execution and destruction of properties (HRW 1995). The Olusegun Obasanjo civilian government attempted national reconciliation by establishing a Truth Commission, but the body's report was never released. Again, the Obasanjo administration failed in its attempt, genuinely or not, to reconcile Ogoni and Shell through the Father Kukah Committee. Shell Oil Company remains, despite all efforts to the contrary, unwelcomed in Ogoni The government seems to have made little other concessions to MOSOP. The militant movement adopted a violently disruptive strategy, and the State characteristically deployed its military might against them. As it did so the costs of disruption on the part of the State, oil companies and militants and innocent civilians escalated. The Musa Yar Adua government was subsequently forced to declare unilateral amnesty for the militants. The differing outcomes suggest that disruptive tactics alone are not sufficient to explain movement outcome

The outcome of disruptive protest varies across places, therefore, I contend that geography is a crucial element in why disruptive action in one place is successful and in another location ends in fiasco. Terrain, which provided 'safe spaces' for militant groups to inflict serial damage on oil infrastructure and by extension, Nigeria's economy, and to hide from reprisal attacks, is a signal difference between militant movements and MOSOP. At will, the State could easily access, occupy the flat and open Ogoni terrain, and visit Ogoni leaders and activists with violence Ogoni leaders with violence; it could not do so in the case of the militants despite the massive deployment of ground, water and air troops. This is taken as the major determinant of the varying outcome of both movements. The critical element was the place where disruptive activities happened.

10 Conclusion

In responding to the research problem of this book, I argued that there is meagre if any gain from the common tendency in the academic debate to explain the Delta conflicts in either materialist or provincial terms. I attempted to utilise issues found in the dominant literature, which portrays the Niger Delta conflict as the legitimate action of local communities. Alternatively, the conflict is seen as no more than a struggle for inclusion in a patronage network propelled by the desire for personal accumulation. Despite the polarisation of approaches to the conflicts, both perspectives share a materialist understanding of conflict that is at best provincial and at worst egocentric. The tendency is a reflection of the proclivity to see conflict-entrepreneurs as self-interested actors, engaged in privileging particularistic interests, while being least concerned with the wellbeing of the larger society.

Reno (2002, 2005) and Omeje (2006) frame the Ogoni movement as a provincial endeavour geared toward the acquisition of material sectional benefits. Others like Osaghae (1995a), Obi (2002), Ikelegbe (2001) and Ibeanu (1997) argue that well-founded concerns about marginalisation, environmental pollution and Shell's insensitivity explain the Ogoni choice of collective action. *A priori* sectional interests alone cannot be solely responsible for, or a motivating factor behind, conflicts in the Niger Delta. The conflict invokes and is equally invoked by communal symbols. It should be a matter for empirical investigation whether provincial economic, political or nationalist interests or, a combination of forms creates the underbelly of the conflicts.

This book focused on why and how the Ogoni conflicts emerged, and engaged with data sourced from written materials, interviews and a questionnaire. Empirical materials considered indicated that MOSOP appealed to provincial demands and others that cannot be reduced to self-interest. In various texts, MOSOP pointed at a skewed federal structure as the basis of ethnic marginalisation, and proposed ethnic autonomy and equality as its resolution. Although the proposed resolution aimed at undoing Ogoni marginalisation, such strategy cannot be conceived as solely self-interested as its anticipated benefits cut across ethnic boundaries and is rationalised in terms of national values and progress. Data from the OBR and interviews with activists indicate that provincial and economic, and nationalistic and symbolic elements composed explanations of why they mobilised. The materialist and provincial dimensions of motivation are evident in Ogoni demand for greater control of oil resources for Ogoni development. At the same time, MOSOP appealed to the symbolic when they accused the ruling elites of betraying true national values. To the Ogoni, Nigeria cannot make progress unless the ethnic minorities were granted equal status with the dominant ethnic groups, couching its claims in the language of justice, fairness and rights. Empirical data detailed both material and symbolic considerations implicated in the decision to mobilise.

The book shows that the Ogoni social conflict was a complex of varying conflicts. The OBR contained demands that amounted to the competitive pursuit of ethnic interests within Nigeria, reform of existing rules or systems, and others that questioned the logic of social organisation. The first category of demands is provincial and material. The second and third categories are not particularistic as the first. They question the rules and value basis of social organisation, even though not necessarily averse to material self-interest. The third category of demands breaches the system's limit of compatibility by delegitimising the existing system and advancing alternative mode of organisation. In other words, the OBR demanded reforms or change of rules in certain areas and systemic change in others.

In addition, the book considers whether individual self-interest alone motivated activists' participation. Empirical data showed that some activists appealed to material needs and personal benefit, while Ogoni leaders emphasised moral motivation. Activists' self-understanding, however, employed a language that made it difficult to compartmentalise their motivation as either material and selfish or symbolic and non self-directed. In such discourses, the materialist and selfish is garnished with the symbolic and non self-directed, and vice versa. For instance, while interested in provincial matters, activists located participation in the movement in the voice of the spirit or spiritual commission or the realisation of being cheated. However, the isolation of motivations served heuristic purposes only because in the discourses and worldview of the Ogoni there was little indication of such separation. For instance, while activists premised mobilisation on environmental pollution, ending the latter had symbolic significance as the land was god.

Arising from a tendency to view the conflict in material and provincial terms, some writers misconstrue the Ogoni conflict as mere reaction to structural disequlibrium. The conflict is conveyed as over-determined by political crisis or structural dislocation. The catalyzing role of culture is discounted. This book argues that contrary to the dominant explanation, Ogoni conflict constitutes a cultural challenge, in its own right, mediated by the active reinterpretation of discourses of environment, national values, federalism and full citizenship. If culture is taken as people's ideas about how social systems and institutions work or should work, then ascendance of new ideas, or perception of incongruence between schemas previously seen as congruent can instigate conflict. The idea of Nigerian federalism has remained unsettled and contested. Ogoni reinterpretation of Nigeria's federalism as lopsided and as 'internal colonialism' suggests continuity between Ogoni challenge and its institutional context. Similarly, cultural organisation of Ogoni was unsettled and contested. Historically, Ogoni was organised based on Yaa tradition, wherein every Ogoni participated in the affairs of the community. The tradition came under assault following the emergence of Ogoni foremost cultural organisation, KAGOTE, which rested on elites control and silencing of women and youth. The inclusion of youths and women in MOSOP, following available historical data on inclusive organisation challenged KAGOTE, reflecting the unsettled and contested nature of culture. Thus, the Ogoni conflict and nature of its mobilisation is properly seen as a cultural challenge traceable to the unsettled and contested relation between ancient and more modern forms of organisation, rather than political crisis. The book further engaged with how the Ogoni mobilised. It departs from existing approaches that made a linear link between grievances and conflict, arguing that they hide micro-mobilisation processes involved in collective mobilisation. This book argued that particularities of place shape the emergence of conflict, its dynamics and trajectories. The reasons people mobilise, what they mobilise for and the spatial dimensions of such dynamics demand empirical investigation. Evidence suggests that experiences of pollution, marginalisation and poverty were not sufficient in themselves to mobilise the Ogoni for contentious action. Activists had to define the condition of Ogoni as unacceptable, that it can be changed, and why and how it should be changed. They created a sense of existential threat and the urgent need to act to save themselves. Ogoni leaders engaged in a redefinition of collective identity and other framing activities, disseminating their ideas through formal and informal meetings. Activists evolved frames that resonated with the people, including the oppressive order and *miideekor* frames. Awareness creation and mobilisation of emotions through such frames galvanised the people despite high cost of participation.

There were other factors, such as the formation of a movement organisation (MOSOP), leadership and incentive, which were critical factors in mobilisation. By building a democratic and inclusive movement organisation, the leaders ensured that the collective aspiration rather than elites' desire propelled MOSOP. Material incentives played a role in mobilisation, as the Ogoni believed success would bring material benefits. However, moral incentives were equally important. Such incentives included the creation of space for ordinary Ogoni to participate in decisions. Ogoni leaders appealed to a glorious Ogoni past and achievements of their ancestors, calling on their age-old courage to confront the present predicament. Such would afford them the opportunity to reshape policies at the national level and assure themselves more negotiating power. The leaders, thus, held out personal sacrifices and risks as a promise, or moral incentive, to improve Ogoni material wellbeing in the future. Similarly, there is empirical evidence that the Ogoni believed that their gods were involved in the struggle, which in turn mobilised their passion and commitment.

There is evidence that collective identity shaped mobilisation and motivated individual participation. This book, thus, suggests that selective incentive is not the only antidote to the free-rider problem. Data suggest that participants did not require material incentives to commit. Some activists refer to other-worldly inspiration or incentives or the moral authority of Saro-Wiwa or anger at being cheated. Thus, for some activists, participation in MOSOP rested on moral incentive, which though non-material, nevertheless harboured the promise of material benefits. Moreover, MOSOP's organisational choices, tactics and strategies were not simply determined by rational calculation of costs and benefits. Rather, identity, democratic values, cultural repertoires, syncretistic religious environment, and belief in the involvement of Ogoni deities shaped the movement and its *modus operandi*. In effect, the dualism between material and symbolic incentives is misleading, as the presence of one does not eliminate the other. Again, this book argues that instrumental logic does not exclude identity concerns and vice versa.

This book raised the question of what role material and self-directed incentives played in mobilising collective action. It provides evidence that in the Ogoni worldview, there is no separation between redistribution and recognition and between the symbolic and materialist. In their claims, activists did not isolate personal benefits from moral considerations. Activists demonstrated little awareness that in their discourses on why they joined the movement, they betrayed personal and non-personal motivations. Evidently, self/other dualism, while useful is misleading because the categories intertwine. The concept, miideekor, emphasises that a phenomenon that is material at one level or perspective may be moral on another level. Ideas saturate human actions. Even the most self-centred idea may orient to moral rules about how to pursue individualistic ends or what to do with it once achieved. Thus, activists' descriptions of their benefits show how they have become more responsible, or what they would never have been without the struggle.

Consideration of why and how Ogoni mobilised centres attention on how spatial factors mediate collective action at multiple levels. Without such sensitivity one would not know why action emerges in one place and not another. This book details how forces across spatial scales altered the Ogoni topography in ways that despoiled the local economy and landscape as well as imperil Ogoni sense of attachment to place. The Ogoni sought to reclaim control of their environment and protect the ghosts that inhabit it. The book provides evidence that in mobilising, Ogoni drew on place-specific features such as beliefs in the *Wiayor*, and involvement of Ogoni spirit and deities, the flat terrain, cultural repertoires of inclusive organisation, the history of Ogoni marginalisation and environmental degradation. Such factors shaped Ogoni frames, strategies, and why and how they mobilised, and gave the movement its unique characteristics.

This book provides evidence that oil extraction as development became conflictual in Ogoni because it penalised the majority of Ogoni while generating immense wealth for the State and Shell. Such development occasioned land theft, destruction of the local economy and livelihoods, and imposed intolerable costs on the Ogoni. Ogoni demands for redress were regularly ignored. The ensuing conflict turned on the struggle to control the mode of development and its governing rules, rather than mere demand for inclusion. The Ogoni struggle was, therefore, development induced rather than a reaction external to it. Such development is clearly undesirable, imperilling community wellbeing, environmental sustainability and national peace. Although the majority of Ogoni experienced development as a penalising phenomenon, their struggle was not a rejection of development but a reappropriation of alternative environmental discourses, sustainable self-development and the right to assert a claim to participative development, which respects Ogoni environment, moral sensibilities and dignity.

Similar collective actions dot the global political landscape. Some of these conflicts are routinely tagged resource conflicts, and their protagonists as greedy and motivated by selfish and/or provincial interests. This book argues that the dualism of selfish and moral motivations is misleading because activists do not make such separation, and theoretical evidence suggests the mutual imbrications of both dimensions. This book suggests that the Ogoni as collective actors embodied virtues that promote national values and while these may appear threatening, a perspective of openness shows their virtue. Their case illustrates that conflict within a resource-rich domain is never merely about resources or environment because in the understanding of the people the environment is 'economic, and it is social and political life and cultural sustenance' (Banks 2002: 42). In that regard, Salih (1999) rightly observes that African environmental politics transcends environment. Serious engagements with collective actions need to dispense with essentialist modernist labels and appreciate the complex worldview of local people in developing societies.

In short, any apprehension of the Ogoni conflict, and similar collective actions, in terms that portray it as only self-oriented is grossly inadequate. Similarly, supposedly corrective reactions, which emphasise genuine grievances, remain equally less than adequate. They do not direct attention to the symbolic, other-directed, and nationalistic aspects of the struggles of the less-powerful. Premised on the *a priori* understanding of collective action as entirely provincial or self-oriented, either perspective fails to further theoretical understanding of reality. In effect, both perspectives trump attempts to understand why poor people risk life and limb in an effort to engender change. Wittingly or unwittingly, they legitimise the status quo while silencing the voices and aspirations of the marginalised. To the extent that this book presents an alternative story, it takes a small first step in staking the nationalist and developmentalist visions embodied by grassroots collective actors, namely the Ogoni.

Appendix

Ogoni Bill of Rights presented to the government and people of Nigeria October, 1990

We, the people of Ogoni (Babbe, Gokana, Ken Khana, Nyo Khana and Tai) numbering about 500,000 being a separate and distinct ethnic nationality within the Federal Republic of Nigeria, wish to draw the attention of the Governments and people of Nigeria to the undermentioned facts:

1 That the Ogoni people, before the advent of British colonialism, were not conquered or colonized by any other ethnic group in present-day Nigeria.
2 That British colonization forced us into the administrative division of Opobo from 1908 to 1947.
3 That we protested against this forced union until the Ogoni Native Authority was created in 1947 and placed under the then Rivers Province.
4 That in 1951 we were forcibly included in the Eastern Region of Nigeria where we suffered utter neglect.
5 That we protested against this neglect by voting against the party in power in the Region in 1957, and against the forced union by testimony before the Willink Commission of Inquiry into Minority Fears in 1958.
6 That this protest led to the inclusion of our nationality in Rivers State in 1967, which State consists of several ethnic nationalities with differing cultures, languages and aspirations.
7 That oil was struck and produced in commercial quantities on our land in 1958 at K. Dere (Bomu oilfield).
8 That oil has been mined on our land since 1958 to this day from the following oilfields: (i) Bomu (ii) Bodo West (iii) Tai (iv) Korokoro (v) Yorla (vi) Lubara Creek and (vii) Afam by Shell Petroleum Development Company (Nigeria) Limited.
9 That in over 30 years of oil mining, the Ogoni nationality have provided the Nigerian nation with a total revenue estimated at over 40 billion Naira (N40 billion) or 30 billion dollars.
10 That in return for the above contribution, the Ogoni people have received NOTHING.

11 That today, the Ogoni people have:

    i   No representation whatsoever in ALL institutions of the Federal Government of Nigeria.

    ii   No pipe-borne water.

    iii No electricity.

    iv   No job opportunities for the citizens in Federal, State, public sector or private sector companies.

    v   No social or economic project of the Federal Government.

12 That the Ogoni languages of Gokana and Khana are underdeveloped and are about to disappear, whereas other Nigerian languages are being forced on us.

13 That the Ethnic policies of successive Federal and State Governments are gradually pushing the Ogoni people to slavery and possible extinction.

14 That the Shell Petroleum Development Company of Nigeria Limited does not employ Ogoni people at a meaningful or any level at all, in defiance of the Federal government's regulations.

15 That the search for oil has caused severe land and food shortages in Ogoni one of the most densely populated areas of Africa (average: 1,500 per square mile; national average: 300 per square mile).

16 That neglectful environmental pollution laws and substandard inspection techniques of the Federal authorities have led to the complete degradation of the Ogoni environment, turning our homeland into an ecological disaster.

17 That the Ogoni people lack education, health and other social facilities.

18 That it is intolerable that one of the richest areas of Nigeria should wallow in abject poverty and destitution.

19 That successive Federal administrations have trampled on every minority right enshrined in the Nigerian Constitution to the detriment of the Ogoni and have by administrative structuring and other noxious acts transferred Ogoni wealth exclusively to other parts of the Republic.

20 That the Ogoni people wish to manage their own affairs.

Now, therefore, while reaffirming our wish to remain a part of the Federal Republic of Nigeria, we make demand upon the Republic as follows:

That the Ogoni people be granted POLITICAL AUTONOMY to participate in the affairs of the Republic as a distinct and separate unit by whatever name called, provided that this Autonomy guarantees the following:

a   Political control of Ogoni affairs by Ogoni people.

b   The right to the control and use of a fair proportion of Ogoni economic resources for Ogoni development.

c   Adequate and direct representation as of right in all Nigerian national institutions.

d   The use and development of Ogoni languages in all Nigerian territory.

e   The full development of Ogoni culture.
f   The right to religious freedom.
g   The right to protect the Ogoni environment and ecology from further degradation.

We make the above demand in the knowledge that it does not deny any other ethnic group in the Nigerian Federation of their rights and that it can only conduce to peace, justice and fair play and hence stability and progress in the Nigerian nation.

We make the demand in the belief that, as Obafemi Awolowo has written: In a true federation, each ethnic group no matter how small, is entitled to the same treatment as any other ethnic group, no matter how large.

We demand these rights as equal members of the Nigerian Federation who contribute and have contributed to the growth of the Federation and have a right to expect full returns from that Federation.

Adopted by general acclaim of the Ogoni people on the 26th day of August, 1990 at Bori, Rivers State and signed by: (see under)

Addendum to the Ogoni Bill of Rights

We, the people of Ogoni, being a separate and distinct ethnic nationality within the Federal Republic of Nigeria, hereby state as follows:

a   That on October 2, 1990 we addressed an Ogoni Bill of Rights to the President of the Federal Republic of Nigeria, General Ibrahim Babangida and members of the Armed Forces Ruling Council;
b   That after a one-year wait, the President has been unable to grant us the audience which we sought to have with him in order to discuss the legitimate demands contained in the Ogoni Bill of Rights;
c   That our demands as outlined in the Ogoni Bill of Rights are legitimate, just and our inalienable right and in accord with civilized values worldwide;
d   That the Government of the Federal Republic has continued, since October 2, 1990, to decree measures and implement policies which further marginalize the Ogoni people, denying us political autonomy, our rights to our resources, to the development of our languages and culture, to adequate representation as of right in all Nigerian national institutions and to the protection of our environment and ecology from further degradation;
e   That we cannot sit idly by while we are, as a people, dehumanized and slowly exterminated and driven to extinction even as our rich resources are siphoned off to the exclusive comfort and improvement of other Nigerian communities, and the shareholders of multi-national oil companies.

NOW, therefore, while re-affirming our wish to remain a part of the Federal Republic of Nigeria, we hereby authorize the Movement for the Survival of Ogoni People (MOSOP) to make representation, for as long as these injustices continue, to the United Nations Commission on Human Rights, the Commonwealth Secretariat, the African Commission on Human and Peoples rights, the European Community and all international bodies which have a role to play in the preservation of our nationality, as follows:

1   That the Government of the Federal Republic of Nigeria has, in utter disregard and contempt for human rights, since independence in 1960 till date, denied us our political rights to self-determination, economic rights to our resources, cultural rights to the development of our languages and culture, and social rights to education, health and adequate housing and to representation as of right in national institutions.

2   That, in particular, the Federal Republic of Nigeria has refused to pay us oil royalties and mining rents amounting to an estimated 20 billion US dollars for petroleum mined from our soil for over thirty-three years.

3   That the Constitution of the Federal Republic of Nigeria does not protect any of our rights whatsoever as an ethnic minority of 500,000 in a nation of about 100 million people and that the voting power and military might of the majority ethnic groups have been used remorselessly against us at every point in time.

4   That multi-national oil companies, namely Shell (Dutch/British) and Chevron (American) have severally and jointly devastated our environment and ecology, having flared gas in our villages for 33 years and caused oil spillages, blow-outs etc., and have dehumanized our people, denying them employment and those benefits which industrial organizations in Europe and America routinely contribute to their areas of operation.

5   That the Nigerian elite (bureaucratic, military, industrial and academic) have turned a blind eye and a deaf ear to these acts of dehumanization by the ethnic majority and have colluded with all the agents of destruction aimed at us.

6   That we cannot seek restitution in the courts of law in Nigeria as the act of expropriation of our rights and resources has been institutionalized in the 1979 and 1989 Constitutions of the Federal Republic of Nigeria, which Constitutions were acts of a Constituent Assembly imposed by a military regime and do not, in any way, protect minority rights or bear resemblance to the tacit agreement made at Nigerian independence.

7   That the Ogoni people abjure violence in their just struggle for their rights within the Federal Republic of Nigeria but will, through every lawful means, and for as long as is necessary, fight for social justice and equity for themselves and their progeny, and in particular demand political autonomy as a distinct and separate unit within the Nigerian nation with full right to (i) control Ogoni political affairs; (ii) use at least fifty per cent of Ogoni economic resources for Ogoni development; (iii) protect the Ogoni environment and ecology from further degradation; and (iv) ensure the full restitution of the harm done to the health of our people

by the flaring of gas, oil spillages, oil blow-outs, etc. by the following oil companies: Shell, Chevron and their Nigerian accomplices.

8   That without the intervention of the international community the Government of the Federal Republic of Nigeria and the ethnic majority will continue these noxious policies until the Ogoni people are obliterated from the face of the earth.

Adopted by general acclaim of the Ogoni people on the 26th day of August 1991 at Bori, Rivers State of Nigeria.

Signed on behalf of the Ogoni people by:

BABBE:

HRH Mark Tsaro-Igbara, Gbenemene Babbe; HRH F.M.K. Noryaa, Menebua, Ka-Babbe; Chief M.A.M. Tornwe III, JP; Prince J.S. Sangha; Dr Israel Kue; Chief A.M.N. Gua.

GOKANA:

HRH James P. Bagia, Gberesako XI, Gberemene Gokana; Chief E.N. Kobani, JP Tonsimene Gokana; Dr B.N. Birabi; Chief Kemte Giadom, JP; Chief S.N. Orage.

KEN-KHANA:

HRH M.H.S. Eguru, Gbenemene Ken-Khane; HRH C.B.S. Nwikina, Emah III, Menebua Bom; Mr. M.C. Daanwii; Chief T.N. Nwieke; Mr. Ken Saro-wiwa; Mr. Simeon Idemyor.

NYO-KHANA:

HRH W.Z.P. Nzidee, Genemene Baa I of Nyo-Khana; Dr G.B. Leton, OON, JP; Mr. Lekue Lah-Loolo; Mr. L.E. Mwara; Chief E.A. Apenu; Pastor M.P. Maeba. TAI: HRH B.A. Mballey, Gbenemene Tai; HRH G.N. Gininwa, Menebua Tua Tua; Chief J.S. Agbara; Chief D.J.K. Kumbe; Chief Fred Gwezia; HRH A. Demor-Kanni, Meneba Nonwa.

References

Adams, W.M. (1996) *Future Nature: A Vision for Conservation*. London: Earthscan.

Adams, W.M. (2003) 'Nature and the Colonial Mind', in William M. Adams and Martin Mulligan (eds), *Decolonizing Nature: Strategies for Conservation in a Post-Colonial Era*, pp.16–50. London: Earthscan.

Adler, M. (2012) 'Collective Identity Formation and Collective Action Framing in a Mexican "Movement of Movements"', *Interface* 4(1): 287–315.

Agbo, A. (2008) 'Oloibiri: Face of the Coming Holocaust', *Tell Magazine* (Special Edition, 18 February), 48–50. Ikeja, Lagos: Tell.

Agbonifo, J. (2003) *The 'Marginalized Violent' Internal Conflict (MVIC) Model and the Ogoni Conflict in the Niger Delta of Nigeria* (American University in Cairo, Egypt, Unpublished MA Thesis, 2003).

Agbonifo, J. (2009) *Development as Conflict: Ogoni Movement, the State and Oil Resources in the Niger Delta, Nigeria*. Published Doctoral Dissertation, Institute of Social Studies, The Hague. Maastricht: Shaker BV.

Aghalino, S.O. (1998) 'British Colonial Policies and the Oil-Palm Industry in the Niger Delta', *Journal of the Pakistan Historical Society* 46(3): 51–62.

Agnew, J. (1987) *Place and Politics: the Geographical Mediation of State and Society*. London: Allen and Unwin.

Agnew, J. (2005) 'Space: Place', in Paul Cloke and Ron Johnston (eds), *Spaces of Geographical Thought*, pp.81–96. London: Sage Publications.

Ajayi, J.F.A. and A.E. Ekoko (1988) 'Transfer of Power in Nigeria: Its Origins and Consequences', Prosser Gifford and William Roger Louis (eds), *Decolonization and African Independence: the Transfer of Power, 1960–1980*, pp.245–269. New Haven and London: Yale University Press.

Ake, C. (1973) 'Explaining Political Instability in New States', *The Journal of Modern African Studies* 11(3): 347–359.

Ake, C. (1996) 'Shelling Nigeria Ablaze' *Tell* vol 129 (January, 29). Lagos: Tell.

Alagoa, E.J. (1970) 'Long-Distance Trade and States in the Niger Delta', *The Journal of African History* 11(3): 319–329.

Alagoa, E.J. (1971) 'The Niger Delta States and their Neighbours, 1600–1800', in J.F. Ade Ajayi and Michael Crowder (eds), *History of West Africa: Volume One*, pp.269–304. London: Longman Group Limited.

Albrecht, S.L., R.G. Amey and S. Amir (1996) 'The Siting of Radioactive Waste Facilities: What are the Effects on Communities?', *Rural Sociology* 61(4): 649–673.

Alexander, J., and P. Smith (1993) 'The Discourse of American Civil Society: A New Proposal for Cultural Studies', *Theory and Society*, 22, 151–207.

Alonale-Laka, D. (2002) 'Shell and Miideekor', *Ogoni Star* 3(1): 13–27.

Amanyie, V. (2001) *The Agony of the Ogonis in the Niger Delta*. Bori, Port Harcourt: Fredsbary Printers & Publishers.

Aminzade, R. and E.J. Perry (2001) 'The Sacred, Religious, and Secular in Contentious Politics: Blurring Boundaries', in R.R. Aminzade, J.A. Goldstone, D. McAdam, E.J. Perry, W.H. Sewell, Jr, S. Tarrow, and C. Tilley (eds), *Silence and Voice in the Study of Contentious Politics*, pp.155–178. New York, NY: Cambridge University Press.

Anderson, M.G. and P.M. Peek (2002) 'Introduction: Charting a Course', in Martha G. Anderson and Philip M. Peek (eds), *Ways of the Rivers: Arts and Environment of the Niger Delta*, pp.25–35. Los Angeles, CA: UCLA Fowler Museum of Cultural History.

Anene, J.C. (1966) *Southern Nigeria in Transition 1885–1906: Theory and Practice in a Colonial Protectorate*. Cambridge: Cambridge University Press.

Anikpo, M. (2002) 'Social Structure and the National Question in Nigeria', in A. Momoh and S. Abejumobi (eds), *The National Question in Nigeria: Comparative Perspectives*, pp.49–67. Hampshire: Ashgate.

Apena, A. (1997) *Colonization, Commerce, and Entrepreneurship in Nigeria: the Western Delta, 1914–1960*. New York: Peter Lang.

Apter, D.E. (1993) 'Democracy, Violence and Emancipatory Movements: Notes for a Theory of Inversionary Discourse', Discussion Paper No. 44, UNRISD (May), 1–43.

Banjo, W.S. (1998) *Oil and Intra-Ethnic Violence in South-Eastern Nigeria: the Internationalisation of Ogoni Crisis*. Lagos: African Research Bureau.

Banks, G. (2002) 'Mining and the Environment in Melanesia: Contemporary Debates Reviewed', *The Contemporary Pacific* 14(1): 39–67.

Bebbington, A. and U. Kothari (2006) 'Transnational Development Networks', *Environment and Planning A* 38: 849–866.

Bechhofer, F. and L. Paterson (2000) *Principles of Research Design in the Social Sciences*. London: Routledge.

Bell, M.M. (1997) 'The Ghosts of Place', *Theory and Society* 26: 813–836.

Benford, R.D. (1993) '"You Could be the Hundredth Monkey": Collective Action Frames and Vocabularies of Motive within the Nuclear Disarmament Movement', *The Sociological Quarterly* 34(2): 195–216.

Benford, R.D. and D.A. Snow (2000) 'Framing Processes and Social Movements: an Overview and Assessment', *Annual Review of Sociology* 26: 611–639.

Berns, M.C. and M.P.N. Roberts (2002) 'Foreword', in Martha G. Anderson and Philip M. Peek (eds), *Ways of the Rivers: Arts and Environment of the Niger Delta*, pp.11–13. Los Angeles, CA: UCLA Fowler Museum of Cultural History.

Bob, C. (2002) 'Political Process Theory and Transnational Movements: Dialectics of Protest among Nigeria's Ogoni Minority', *Social Problems* 49(3): 395–415.

Bob, C. (2005) *The Marketing of Rebellion: Insurgents, Media, and International Activism*. New York: Cambridge University Press.

Boele, R., H. Fabig and D. Wheeler (2001) 'Shell, Nigeria and the Ogoni: a Study in Unsustainable Development', *Sustainable Development* 9: 74–86.

Breines, W. (1983) *Community and Organization in the New Left, 1962–1968: The Great Refusal*. New Brunswick and London: Rutgers University Press.

Brohman, J. (1995) 'Economism and Critical Silences in Development Studies: a Theoretical Critique of Neoliberalism', *Third World Quarterly* 16(2): 297–318.

Catholic Relief Services (2003) *Bottom of the Barrel: Africa's Oil Boom and the Poor.* Baltimore, MD: Catholic Relief Services.

Catholic Secretariat of Nigeria (2006) *Nigeria: the Travesty of Oil and Gas Wealth.* Lagos: Catholic Secretariat.

Chakrabathy, Dipesh (2000) *Provincializing Europe, Postcolonial thought and Historical Difference.* Princeton: Princeton University Press.

Civil Liberties Organization (1996) *Ogoni: Trials and Travails.* Yaba, Lagos: Civil Liberties Organization.

Collier, P. (2001) 'Economic Causes of Civil Conflict and their Implications for Policy', in Chester A. Crocker, Fen Osler Hampson and Pamela Aall (eds), *Turbulent Peace: the Challenges of Managing International Conflict.* Washington DC: United States Institute for Peace Press.

Collier, P. and A. Hoeffler (2000) 'Greed and Grievance in Civil War', *Policy Research Working Paper 2355.* Washington DC: World Bank.

Collier, Paul and A. Hoeffler (2002) 'Greed and Grievance in Civil War'. World Bank, DECRG. Online at http://econ.worldbank.org/programs/conflict.

Coulson, E. (2009) 'Movement for the Emancipation of the Niger Delta: Political Marginalisation, Repression and Petro-Insurgency', Discussion Paper No. 47, The Nordic African Institute.

Cowen, M.P. and R.W. Shenton (1996) *Doctrines of Development.* London and New York: Routledge.

Cresswell, T. (2004) *Place: A Short Introduction.* Oxford: Blackwell Publishing.

Crowder, M. (1973) *The Story of Nigeria.* London: Faber and Faber.

Dibua, J.I. (2006) *Modernization and the Crisis of Development in Africa: the Nigerian Experience.* Aldershot: Ashgate.

Dikec, M. (2005) 'Space, Politics, and the Political', *Environment and Planning D: Society and Space* 23: 171–188.

McAdam, Doug, John D. McCarthy and Mayer N. Zald, (1996) 'Introduction', In Doug McAdam, John D. McCarthy, and Mayer N. Zald (eds.), *Comparative Perspectives on Social Movements: Political Opportunities, Mobilizing Structures, and Cultural Framings*, pp.1–20. Cambridge: Cambridge University Press.

Doyle, T. (2008) 'The Politics of Hope: Understanding Environmental Justice and Security in the Indian Ocean Region within a Post-Colonialist Frame', in Timothy Doyle and Melissa Risely (eds), *Crucible for Survival: Environmental Security and Justice in the Indian Ocean Region*, pp.305–324. New Brunswick: Rutgers University Press.

Dunning, T. and L. Wirpsa (2004) 'Oil and the Political Economy of Conflict in Colombia and Beyond: a Linkage Approach', *Geopolitics* 14(48): 81–108.

Elias, Norbert (1991) *The Society of Individuals.* Oxford:Basil Blackwell.

Escobar, A. (1984) 'Discourse and Power in Development: Michel Foucault and the Relevance of his Work to the Third World', *Alternatives* 10(3): 377–400.

Escobar, A. (1995) *Encountering Development: Thre Making and Unmaking of the Third World.* Princeton, NJ: Princeton University Press.

Escobar, A. (2003) 'Displacement, Development, and Modernity in the Colombian Pacific', *International Social Science Journal* 175: 157–167.

Fairhead, J. and M. Leach (1995) '*False Forest History, Complicit Social Analysis: Rethinking some West African Environmental Narratives*', *World Development* 23(6): 1023–1035.

Ferguson, J. (1994) *The Anti-Politics Machine: 'Development', Depoliticization, and Bureaucratic Power in Lesotho.* Cambridge: Cambridge University Press.

Ferree, M.M. (1992) 'The Political Context of Rationality: Rational Choice Theory and Resource Mobilization', in Aldon D. Morris and Carol McClurg Mueller (eds), *Frontiers of Social Movement Theory*, pp.29–52. New Haven, CT: Yale University Press.

Finlay, L. (2002). 'Negotiating the Swamp: The Opportunity and Challenge of Reflexivity in Research Practice', *Qualitative Research* 2(2): 209–230.

Fligstein, N. and D. McAdam (1995) 'A Political-Cultural Approach to the Problem of Strategic Action'. Available online at http://sociology.berkeley.edu/profiles/fligstein/pdf/DOGPAP.03.pdf

Foster, George (1962) *Traditional Cultures*. New York: Harper and Brothers.

Freudenburg, W.R. (2005) 'Privileged Access, Privileged Accounts: Towards a Socially Structured Theory of Resources and Discourses', *Social Forces* 84(1): 89–114.

Frynas, J.G. (2000) *Oil in Nigeria: Conflict and Litigation between Oil Companies and Village Communities*. Hamburg: LIT.

Gaventa, J. (1982) *Power and Powerlessness: Quiescence and Rebellion in an Appalachian Valley*. Illinois: University of Illinois Press.

GerhardsJ., and D. Rucht (1992) 'Mesomobilization: Organizing and Framing in Two Protest Campaigns in West Germany', *American Journal of Sociology* 98(3): 555–595.

Gertzel, C. (1962) 'Relations between African and European Traders in the Niger Delta 1880–1896', *The Journal of African History* 3(2): 361–366.

Goulet, D.A. (1987) 'Participation in Development Decisions as Moral Incentive: a Brazilian Case Study', *Development and Peace* 8(1): 131–137.

Goulet, D.A. (1968) 'Development for What?', *Comparative Political Studies* 1(2): 295–312.

Goulet, D.A. (1980) 'Development Experts: the One-Eyed Giants', *World Development* 8: 481–489.

Goulet, D.A. (1987) 'Participation in Development Decisions as Moral Incentive: a Brazilian Case Study', *Development and Peace* 8(1): 131–137.

Goulet, D.A. (1992) 'Development: Creator and Destroyer of Values', *World Development* 20(3): 467–475.

Goulet, D.A. and C.K. Wilber (1996) 'The Human Dilemma of Development', in Kenneth P. Jameson and Charles K. Wilber (eds), *The Political Economy of Development and Underdevelopment* (6th edition), pp.469–476. New York: McGraw-Hill, Inc.

Grillo, R.D. (1997) 'Discourses of Development: the View of Anthropology', in R.D. Grillo and R.L. Stirrat (eds), *Discourses of Development: Anthropological Perspectives*, pp.1–33. Oxford: Berg.

Gurr, T.R. (1970) *Why Men Rebel*. Princeton, NJ: Princeton University Press.

Habermas, J. (1987) *The Theory of Communicative Action* (2nd edition). Boston: Beacon Press.

Haynes, J. (2006) 'Political Critique in Nigerian Video Films', *African Affairs* 105(421): 511–533.

Henry, L., G. Mohan and H. Yanacopulos (2004) 'Networks as Transnational Agents of Development', *Third World Quarterly* 25(5): 839–855.

Homer-Dixon, T. (1999) *Environment, Scarcity, and Violence*. Princeton, London: Princeton University Press.

Hopkins, A.G. (1968) 'Economic Imperialism in West Africa: Lagos, 1880–1892', *The Economic History Review*, New Series 21(3): 580–606.

Hulme, D. and M. Turner (1990) *Sociology and Development: Policies, and Practices.* New York: St. Martin's Press.

Human Rights Watch (1995) 'Nigeria-The Ogoni Crisis: A Case-Study of Military Repression in Southeast Nigeria', 7(5). Human Rights Watch/Africa. Available at www.hrw.org/report/1995/07/01/ogoni-crisis/case-study-military-repression-southea stern-nigeria

Ibeanu, O. (1997) 'Oil, Conflict and Security in Rural Nigeria: Issues in Ogoni Crisis', African Association of Political Science, *Occasional Papers Series* 1(2).

Ibeanu, O. (2000) 'Oiling the Friction: Environmental Conflict Management in the Niger Delta, Nigeria', *Environmental Change and Security Project Report* 6: 19–32.

Ifemesia, C.C. (1982) 'Nigeria: The Country of the Niger Area', in Boniface I. Obichere (ed.), *Studies in Southern Nigerian History*, pp.21–36. London: Frank Cass and Company Limited.

Ikelegbe, A. (2001) 'Civil Society, Oil and Conflict in the Niger Delta Region of Nigeria: Ramifications of Civil Society for a Regional Resource Struggle', *Journal of Modern African Studies* 39(3): 437–469.

Isumonah, V.A. (2004) 'The Making of the Ogoni Ethnic Group', *Africa* 74(3): 433–453.

Jones, G.I. (1963) *The Trading States of the Oil Rivers: A Study of Political Development in Eastern Nigeria.* London: Oxford University Press.

Jung, C. (2003) 'The Politics of Indigenous Identity: Neoliberalism, Cultural Rights, and the Mexican Zapatistas', *Social Research* 70(2): 433–462.

Klandermas, B., J.M. Sabucedo, M. Rodriguez and M. de Weerd (2002) 'Identity Processes in Collective Action Participation: Farmers' Identity and Farmers' Protest in the Netherlands and Space', *Political Psychology* 23(2): 235–251.

Koenig-Archibugi, M. (2004) 'Transnational Corporations and Public Accountability', *Government and Opposition* 39(2): 234–259.

Johnston, Hank and Bert Klandermans (eds) (1995) *Social Movements and Culture.* Minneapolis, MN: University of Minnesota Press.

Kpone-Tonwe, S. (1997) 'Property Reckoning and Methods of Accumulating Wealth among the Ogoni of the Eastern Niger Delta', *Africa* 67(1): 130–158.

Kpone-Tonwe, S. (2003) *Youth and Leadership: Training in the Niger Delta the Ogoni Example.* Port Harcourt: Onyoma Research Publications.

Kpone-Tonwe, S. and J. Salmons (2002) 'The Arts of the Ogoni', in M.G. Anderson and P.M. Peek (eds), *Ways of the Rivers: Arts and Environment of the Niger Delta*, pp.275–301. Los Angeles, CA: UCLA.

Laclau, Ernesto and Chantal Mouffe. 1985. *Hegemony and Socialist Strategy.* London: Verso.

Lachmann, R. and N.A. Pichardo (1994) 'Making History from Above and Below: Elite and Popular Perspectives on Politics', *Social Science History* 18(4): 497–504.

Lefebvre, H. (1991) *The Production of Space.* Oxford: Blackwell.

Leftwich, A. (2000) *States of Development: On the Primacy of Politics in Development.* Cambridge: Polity Press.

Leis, Philip (1964) 'Palm Oil, Illicit Gin, and the Moral Order of the Ijaw', *American Anthropologist* 66: 828–838.

Leis, P.E. (2002) 'Preface: Cultural Identity in the Multicultural Niger Delta', in Martha G. Anderson and Philip M. Peek (eds), *Ways of the Rivers: Arts and Environment of the Niger Delta*, pp.15–21. Los Angeles, CA: UCLA Fowler Museum of Cultural History.

Livesey, S.M. (2001) 'Eco-identity as Discursive Struggle: Royal/Dutch Shell, Brent Spar, and Nigeria', *Journal of Business Communication* 38(1): 58–91.

Lockie, Stewart (2004), 'Collective Agency, Non-Human Causality and Environmental Social Movements: A Case Study of the Australian "Landcare Movement"', *Journal of Sociology* 40(1): 41–57.

Loolo, G.N. (1981). *A History of Ogoni*. Port Harcourt: N.P.

Loveman, M. (2005) 'The Modern State and the Primitive Accumulation of Symbolic Power', *AJS* 110(6): 1651–1683.

Maier, K. (2000) *This House Has Fallen: Nigeria in Crisis*. London: Penguin Books.

Martin, D.G. and Miller, B. (2003) 'Space and Contentious Politics', *Mobilization* 8(2): 143–156.

Marquardt, Erich (2007) 'The Niger Delta Insurgency and Its Threat to Energy Security', *Terrorism Monitor* 4(16). Available at https://jamestown.org/program/the-niger-delta-insurgency-and-its-threat-to-energy-security/

Mbaeyi, P.M. (1982) 'The British Navy and 'Southern Nigeria' in the Nineteenth Century', in Boniface Obichere (ed.), *Studies in Southern Nigerian History*, pp.201–218. London: Frank Cass.

Mbeke-Ekanem, T. (2000) *Beyond the Execution: Understanding the Ethnic and Military Politics in Nigeria*. Los Angeles, CA: Crystal Graphic Communications and Publishing Co.

McAdam, D. (1982) *Political Process and Development of Black Insurgency, 1930–1970*. Chicago: University of Chicago Press.

McAdam, D. (1996) 'Political Opportunities: Conceptual Origins, Current Problems, Future Directions', in Doug McAdam, John D. McCarthy and Mayer N. Zald (eds), *Comparative Perspectives on Social Movements: Political Opportunities, Mobilizing Structures, and Cultural Framings*, pp.23–40. Cambridge: Cambridge University Press.

McAdam, D., J.D. McCarthy and M.N. Zald (1996) 'Introduction', in Doug McAdam, John D. McCarthy and Mayer N. Zald (eds), *Comparative Perspectives on Social Movements: Political Opportunities, Mobilizing Structures, and Cultural Framings*, pp.1–20. Cambridge: Cambridge University Press.

McAdam, D., S. Tarrow and C. Tilly (2001) *Dynamics of Contention*. Cambridge: Cambridge University Press.

McCarthy, J.D. and M.N. Zald (1977) 'Resource Mobilization and Social Movements: a Partial Theory', *American Journal of Sociology* 82: 1212–1241.

Mehretu, A. (1989) *Regional Disparity in Sub-Saharan Africa: Structural Readjustment of Uneven Development*. Boulder, CO: Westview Press.

Melson, R. and H. Wolpe (1970). 'Modernization and the Politics of Communalism: A Theoretical Perspective', *The American Political Science Review* 64(4): 1112–1130.

Melucci, A. (1985) 'The Symbolic Challenge of Contemporary Movements', *Social Research* 52(4): 789–816.

Melucci, A. (1989) *Nomads of the Present: Social Movements and Individual Needs in Contemporary Society*. Philadelphia: Temple University Press.

Melucci, A. (1996) *Challenging Code: Collective Action in the Information Age*. Cambridge: Cambridge University Press.

Mikesell, R.F. et al. (1971) *Foreign Investment in the Petroleum and Mineral Industries: Case Studies of Investor-Host Country Relations*. Baltimore, MD: The Johns Hopkins Press.

Miller, B.A. (1994) *Geography and Social Movements: Comparing Anti-Nuclear Activism in the Boston Area*. Minnesota: University of Minnesota Press.

Mittelman, J.H. (1998) 'Globalization and Environmental Resistance Politics', *Third World Quarterly* 19(5): 847–872.

Momoh, A. (2002) 'The Philosophy and Theory of the National Question', in Abubakar Momoh and Said Adejumobi (eds), *The National Question in Nigeria: Comparative Perspectives*, pp.1–30. Aldershot: Ashgate.

Monroe, K.R. (2001) 'Morality and a Sense of Self: The Importance of Identity and Categorization for Moral Action', *American Journal of Political Science* 45(3): 491–507.

Monroe, K.R., Martin, A. and Ghosh, P. (2009) 'Politics and an Innate Moral Sense: Scientific Evidence for an Old Theory?', *Political Research Quarterly* 62(3): 614–634.

Moore, Barrington, Jr (1978) *Injustice: the Social Bases of Obedience and Revolt.* London: Palgrave Macmillan.

Morris, Aldon (1984). *The Origins of the Civil Rights Movement: Black Communities Organizing for Chunge.* New York: Free Press.

MOSOP (1990) Ogoni Bill of Rights. Available at www.bebor.org/wp-content/uploads/2012/09/Ogoni-Bill-of-Rights.pdf

MOSOP (2004) *Ken Saro-Wiwa's Last Word.* Port Harcourt: MOSOP.

Naanen, B. (1995) 'Oil-Producing Minorities and the Restructuring of Nigerian Federalism: The Case of the Ogoni People', *Journal of Commonwealth and Comparative Politics* 33(1): 46–78.

Nepstad, Sharon Erickson and Bob, Clifford (2006) 'When Do Leaders Matter? Hypotheses on Leadership Dynamics in Social Movements', *Mobilization: An International Journal* 11(1): 1–22.

North, D.C. (1955) 'Location Theory and Regional Economic Growth', *The Journal of Political Economy* LXIII(3): 243–258.

Northrup, D. (1978) *Trade without Rulers: Precolonial Economic Development in South-Eastern Nigeria.* Oxford: Clarendon.

Nwabughuogu, A.I. (1982) 'From Wealthy Entrepreneurs to Petty Traders: the Decline of African Middlemen in Eastern Nigeria, 1900–1950', *Journal of African History* 23(3): 365–379.

Obi, C. (1997) 'Globalisation and Local Resistance: the Case of the Ogoni Versus Shell', *New Political Economy* 2(1): 137–148.

Obi, C. (1999) 'Globalisation and Environmental Conflicts in Africa', *African Journal of Political Science* 4(1): 40–62.

Obi, C. (2002) 'Oil and the Minority Question', *The National Question in Nigeria: Comparative Perspectives* 99–100.

Obi, C. (2006) 'Youth and the Generational Dimensions to Struggles for Resource Control in the Niger Delta: Prospect for the Nation-State Project in Nigeria', *Codesria Monograph Series*, pp.1–48.

Obi, C. (2009) 'Structuring transnational spaces of identity, rights and power in the Niger Delta of Nigeria', *Globalizations* 6(4): 467–481.

Ogoni Charities Inc. Available at www.ogonicharity.camp7.org/Default.aspx?pageId=61652. Accessed 23 August 2012.

Ogoni Star (n.d.) 'Ogoni 9 Families Harmer on Unity, Release of Bodies', 2(11) (Special Edition). Bori, Ogoni: Ogoni Star Communications.

Okafor, F.C. (1980) 'Integrated Rural Development Planning in Nigeria: A Spatial Dimension', *Cahiers d'etudes africaines*, 20: 83–95.

Okilo, Chief Melford (June 2000) cf Ambily Etekpe (2003) 'The Evolution of Chief Dappa-Biriye's Political Thought', in E.J. Alagoa (ed.), *Harold Dappa-Biriye: His*

Contributions to Politics in Nigeria, pp.21–40. Port Harcourt: Onyoma Research Publications.

Okome, O. (ed.) (2000) *Before I am Hanged: Ken Saro-Wiwa, Literature, Politics and Dissent*. Trenton, NJ: Africa World Press.

Okonta, Ike (2008) *When Citizens Revolt: Nigerian Elites, Big Oil, and the Ogoni Struggle for Self-Determination*. Port Harcourt: Ofirima Publishing House Ltd.

Okorobia, A.M. (1999). A History of the Underdevelopment of the Eastern Niger Delta, ad 1500–1993. Unpublished PhD Dissertation, University of Port Harcourt.

Olson, M. 1965. *The Logic of Collective Action: Public Goods and the Theory of Groups*. Cambridge, MA: Harvard University Press.

Omeje, K. (2005) 'Oil Conflict in Nigeria: Contending Issues and Perspectives of the Local Niger Delta People', *New Political Economy* 10(3): 321–334.

Omeje, K. (2006) *High Stakes and Stakeholders: Oil Conflict and Security in Nigeria*. Aldershot: Ashgate.

Omoruyi, O. (2000) 'The Politics of Oil: Who Owns Oil, Nigeria, States or Communities?'. Available online at http://nigeriaworld.com/feature/publication/omoruyi/oil.html.

Organization of African Unity (OAU) (1982) *Lagos Plan of Action for the Economic Development of Africa, 1980–2000, 2nd edition*. Geneva: International Institute for Labour Studies.

Ortner, Sherry B. (1995) 'Resistance and the Problem of Ethnographic Refusal', *Comparative Studies of Society and History* 37(1): 173–193.

Orubu, C.O., A. Odusola and W. Ehwarieme (2004) 'The Nigerian Oil Industry: Environmental Diseconomies, Management Strategies and the Need for Community Involvement', *Journal of Human Ecology* 16(3): 203–214.

Osadolor, O.B. (2002) 'The National Question in Historical Perspective', in Abubakar Momoh and Said Adejumobi (eds), *The National Question in Nigeria: Comparative Perspectives*, pp.31–48. Aldershot: Ashgate.

Osaghae, E.E. (1995a) 'Amoral Politics and Democratic Instability in Africa: a Theoretical Exploration', *Nordic Journal of African Studies* 4(1): 62–78

Osaghae, E.E. (1995b) 'The Ogoni Uprising: Oil Politics, Minority Agitation and the Future of the Nigerian State', *African Affairs* 94(376): 325–344.

Osaghae, E.E. (2008) 'Social Movements and Rights Claims: The Case of Action Groups in the Niger Delta of Nigeria', *Voluntas* 19(2): 189–210.

Osha, S. (2006) 'Birth of the Ogoni Protest Movement', *Journal of Asian and African Studies* 4(1–2): 13–38.

Oslender, U. (2004) 'Fleshing out the Geographies of Social Movements: Colombia's Pacific Coast Black Communities and the "Aquatic Space"', *Political Geography* 23: 957–985.

Osoba, S.O. (1987) 'The Transition to Neocolonialism', in Toyin Falola (ed.), *Britain and Nigeria: Exploitation or Development?*, pp.223–248. London: Zed Books.

Otite, O. (1990) 'Rural Poverty in Nigerian Bourgeois Sociology: a Materialist Critique', in O. Otite and C. Okali (eds), *Readings in Nigerian Rural Society and Rural Economy*. Oxford and Portsmouth, NH:Heinemann Educational Books.

Parfitt, T. (2002) *The End of Development: Modernity, Post-Modernity and Development*. London: Pluto Press.

Payne. Charles (1995) *I've Got the Light of Freedom: The Organizing Tradition and the Mississippi Freedom Struggle*. Berkeley, CA: University of California Press.

Pegg, S. (2000) 'Ken Saro-Wiwa: Assessing the Multiple Legacies of a Literary Interventionist', *Third World Quarterly* 21(4): 701–708.

Peluso, N. and Watts, M. (2001) *Violent Environments.* Ithaca: Cornell University Press.

Pile, S. and M. Keith (1997) *Geographies of Resistance.* London: Routledge.

Pilgrim, Sarah and Jules Pretty (2010) 'Nature and Culture: An Introduction', in Sarah Pilgrim and Jules Pretty (eds) *Nature and Culture: Rebuilding Lost Connections,* pp.1–20. London and Washington, DC: Earthscan.

Piven, Frances Fox and Richard A. Cloward. (1977) *Poor People's Movement: Why They Succeed, How They Fail.* New York: Pantheon Books.

Polletta, Francesca (2008) 'Culture and its Discontents', *Sociological Inquiry* 67(4): 431–450.

Polletta, F. (2008) 'Culture and Movements', Annals, *AAPSS* 69: 78–96.

Polletta, F. and J.M. Jasper (2001) 'Collective Identity and Social Movements', *Annual Review of Sociology* 27(1): 283–305.

Polletta, F. and M.K. Ho (2006) 'Frames and their Consequences' in R.E. Goodin and C. Tilly (eds), *The Oxford Handbook of Contextual Political Analysis,* pp.187–209. Oxford: Oxford University Press.

Preston, P. (2002). *Development Theory: An Introduction.* Oxford: Blackwell Publishers.

Radford, G. (1992). 'Positivism, Foucault, and the Fantasia of the Library: Conceptions of Knowledge and the Modern Library Experience', *Library Quarterly* 62(4): 408–424.

Redfield, Robert (1953) *The Primitive World and its Transformation.* Ithaca, NY: Cornell University Press.

Reno, W. (2000) 'Shadow States and the Political Economy of Civil War', in Mats Berdal and David M. Malone (eds), *Greed and Grievances: Economic Agendas in Civil Wars,* pp.43–68. Boulder, CO: Lynne Rienner.

Reno, W. (2002) 'The Politics of Insurgency in Collapsing States', *Development and Change* 33(5): 837–858.

Reno, W. (2005) 'The Politics of Violent Opposition in Collapsing States', *Government and Opposition* 40(2): 123–327.

Ribeiro, G.L. (2002) 'Power, Networks and Ideology in the Field of Development: New Solutions to Old Problems', in Carlos Lopes, Khalid Malik and Sakiko Fukuda-Parr (eds) *Capacity for Development: New Solutions to Old Problems,* pp.168–184. London: Earthscan.

Salih, M.A. (1999) *Environmental Politics and Liberation in Contemporary Africa.* Dordrecht: Kluwer Academic Publishers.

Saro-Wiwa, K. (1992) *Genocide in Nigeria: the Ogoni Tragedy.* Port Harcourt: Saros International Publishers.

Saro-Wiwa, K. (1995) *A Month and a Day: a Detention Diary.* Ibadan: Spectrum Books Limited.

Schatz, S.P. (1984) 'Pirate Capitalism and the Inert Economy of Nigeria', *The Journal of Modern African Studies* 22(1): 45–57.

Schuurman, F.J. (1993) *Beyond the Impasse: New Directions in Development Theory.* London: Zed Books.

Schwandt, T. (2002). 'Three Epistemological Stances for Qualitative Inquiry', in N. Denzin and Y. Lincoln (eds) *The Landscape of Qualitative Research: Theories and Issues,* pp.189–213. London: Sage Publications.

Scoones, I. and J. Thompson (1993). 'Challenging the Populist Perspective: Rural People's Knowledge, Agricultural Research and Extension Practice', *IDS Discussion Paper No. 332.*

Scott, J.C. (1998) *Seeing Like a State: How Certain Schemes to Improve the Human Condition Have Failed*. New Haven: Yale University Press.

Sen, A. (1999) *Development as Freedom*. Oxford: Oxford University Press.

Sewell, W.A., Jr (1992) 'A Theory of Structure: Duality, Agency, and Transformation', *The American Journal of Sociology* 98(1): 1–29.

Sewell, W.A., Jr (2001) 'Space in Contentious Politics', in R.R. Aminzade, J.A. Goldstone, D. McAdam, E.J. Perry, W.H. Sewell, Jr, S. Tarrow and C. Tilly (eds), *Silence and Voice in the Study of Contentious Politics*, pp.51–88. Cambridge: Cambridge University Press.

Shaw, Timothy M. (2003) 'Conflict and Peace-building in Africa: The Regional Dimensions', Wider Discussion Paper No. 2003/10. United Nations.

Simmons, C.S. (2005) 'Territorialising Land Conflict: Space, Place, and Contentious Politics in the Brazilian Amazon', *Geojournal* 64: 307–317.

Snow, D.A. (2004) 'Framing Processes, Ideology, and Discursive Fields', in David A. Snow, Sarah A. Soule and Hanspeter Kriesi (eds), *The Blackwell Companion to Social Movements*, pp.380–412. Malden, MA: Blackwell Publishing.

Snow, D.A. and R. Benford (1988) 'Ideology, Frame Resonance and Participant Mobilization', *International Social Movement Research* 1: 197–219.

Snow, D.A. and R. Benford (1992) 'Master Frames and Cycles of Protest', in Aldon D. Morris and Carol McClurg Mueller (ed.), *Frontiers of in Social Movement Theory*, pp.133–155. New Haven: Yale University Press.

Snow, D.A., E. Burke Rochford, Jr, Steven K. Worden, and R. Benford (1986) 'Frame Alignment Processes, Micromobilisation and Movement Participation,' *American Sociological Review* 51(4): 464–481.

Starn, O. (1992) '"I Dreamed of Foxes and Hawks": Reflections on Peasant Protest, New Social Movements, and the Rondas Campesinas of Northern Peru', in Arturo Escobar and Sonia E. Alvarez (eds), *The Making of Social Movements in Latin America: Identity, Strategy, and Democracy*, pp.89–111. Boulder, CO: Westview Press.

Stolper, W.F. (1966) *Planning Without Facts: Lessons in Resource Allocation from Nigeria's Development*. Massachusetts: Harvard University Press.

Stryker, S., T.J. Owens and R.W. White (2000) *Self, Identity, and Social Movements*. Minneapolis: University of Minnesota Press.

Swidler, A. (1995) *Cultural Power and Social Movements*, in Hank Johnston and Bert Klandermans (eds) *Social Movements and Culture*, pp.25–41. Minneapolis: University of Minnesota Press.

Swyngedouw, E. (1999) 'Modernity and Hybridity: Nature, Regeneracionismo, and the Production of the Spanish Waterscape, 1890–1930', *Annals of the Association of American Geographers* 89(3): 443–465.

Tamuno, T.N. (1970) 'Separatist Agitations in Nigeria since 1914', *The Journal of Modern African Studies* 8(4): 563–584.

Tamuno, T.N. (1972) *The Evolution of the Nigerian State: The Southern Phase, 1898–1914*. London: Longman.

Tamuno, T.N. (1978) *The Evolution of the Nigerian State: the Southern Phase 1895–1914*. London: Longman.

Tangwa, G. (2004) 'Some African Reflections on Biomedical and Environmental Ethics', in Wiredu Kwasi (ed.) *A Companion to African Philosophy*, pp.387–395. Oxford: Blackwell Publishers.

Tarrow, Sidney (2001) 'Silence and Voice in the Study of Contentious Politics: Introduction', in Ronald Aminzade, Jack A. Goldstone, Doug McAdam, Elizabeth J. Perry,

William H.Sewell, Jr, Sidney Tarrow, Charles Tilley (eds), *Silence and Voice in the Study of Contentious Politics*, pp.1–13. Cambridge: Cambridge University Press.

Taylor, P.J. (1994) 'The State as Container: Territoriality in the Modern World-System', *Progress in Human Geography* 18(2): 151–162.

Teske, N. (1997) 'Beyond Altruism: Identity-Construction as Moral Motive in Political Explanation', *Political Psychology* 18(1): 71–91.

Tilly, C. (1978) *From Mobilization to Revolution*. Reading: Addison-Wesley.

Tilly, C. (2000) 'Spaces of Contention', *Mobilization* 5(2): 135–159.

Touraine, A. (1981) *The Voice and the Eye: An Analysis of Social Movements*. Cambridge: Cambridge University Press.

Touraine, A. (1985) 'An Introduction to the Study of Social Movements', *Social Research* 52(4): 749–787.

Touraine, A. (2000) *Can We Live Together? Equality and Difference*. Cambridge: Polity Press.

Tschirgi, D. (1999) 'Marginalized Violent Internal Conflict in the Age of Globalization: Egypt and Mexico', *Arab Studies Quarterly* 21(3): 13–34.

Tschirgi, D. (2007) *Turning Point: the Arab World's Marginalization and International Security after 9/11*. Westport: Praeger Security International.

UNCTAD (2002) *World Investment Report 2002: TNCs and Export Competitiveness*. New York and Geneva: United Nations.

UNDP (2006) *Niger Delta Human Development Report*. Abuja, Nigeria: UNDP.

UNPO (1995) *Ogoni: Report of the UNPO Mission to Investigate the Situation of the Ogoni of Nigeria February 17–26, 1995*. The Hague: UNPO.

Watts, M. (2004) 'Resource Curse? Governmentality, Oil and Power in the Niger Delta, Nigeria', *Geopolitics* 14(47): 50–80.

Welch, C.E., Jr (1995) 'The Ogoni and Self-Determination: Increasing Violence in Nigeria', *Journal of Modern African Studies* 33(4): 635–649.

Whitt, L.A. and J.D. Slack (1994) 'Communities, Environment and Cultural Studies', *Cultural Studies* 8(1): 5–31.

Wilton, Robert D. and Cynthia Cranford (2002) 'Toward an Understanding of the Spatiality of Social Movements: Labor Organizing at a Private University in Los Angeles', *Social Problems* 49(3): 374–394.

Wolf, E.R. (1969) *Peasant Wars of the Twentieth Century*. New York: Harper and Row.

Wolford, W. (2003) 'Families, Fields, and Fighting for Land: the Spatial Dynamics of Contention in Rural Brazil', *Mobilization* 8(2): 157–172.

World Bank (1981) *Accelerated Development in Sub-Saharan Africa: An Agenda for Action*. Washington DC: The World Bank.

Young, I.M. (1990) 'The Ideal of Community and the Politics of Difference', in Linda J. Nicholson (ed.), *Feminism/Postmodernism*, pp.300–323. New York: Routledge.

Zuckerman, M. (1978) 'Dreams that Men Dare to Dream: the Role of Ideas in Western Modernization', *Social Science History* 2(3): 332–345.

Personal interviews

AA, pseudonym for a former militant, interviewed by phone, May 2011.

Bari-ara, Kpalap, Ogoni activist and publicity secretary of MOSOP, Port Harcourt, 2 March 2006.

Barikor, Dr Innocent, Port Harcourt, 4 December, 2007.

Biragbara, Rev. Richard, personal interview, Port Harcourt, 2012.

Deemua, Chief Denis, Uegwere, 9 March 2008.

Dickson, Charity, female Ogoni activist, trader, MOSOP office, Port Harcourt, 11 March 2008.

Douglas, Oronto, human rights lawyer and activist, Port Harcourt, 6 March 2006.

EK, pseudonym for ex-militant, interviewed by phone, May 2011.

JC, pseudonym for ex-militant, interviewed by phone, May 2011.

Kigbara, Young, personal interview, Port Harcourt, 23 May 2012.

Kpalap, Bari-ara, publicity secretary of MOSOP, Port Harcourt, 20 May 2008.

Kpone-Tonwe, Dr Sonpie, Ogoni historian, pastor and university professor, Port Harcourt, 7 March 2008.

Laka, Des, personal interview, Port Harcourt, 11 March 2008.

Legborsi, Saro Pyagbara, personal interview, Port Harcourt, 2006, 2007 and 2012.

ME, pseudonym for ex-militant, interviewed by phone, May 2011.

Moses, Damgbor, Ogoni activist, MOSOP office, Port Harcourt, 9 December 2008.

Naanen, Ben, personal interviews held in Port Harcourt, May 2006 and 8 December 2008.

Nagbo, Carolyn, female Ogoni activist, MOSOP office, Port Harcourt, 8 December 2008.

Nkalaa, Chief Emmanuel, activist and former chair of the youth arm of MOSOP, personal interviews, Port Harcourt, 5 February 2008 and 22 May 2012.

Nwigani, Roseline, school teacher, female activist and former leader of the women arm of MOSOP, Port Harcourt, 5 February 2008.

Tanee, Wilfred, Ogoni activist, Bori, Ogoni, 8 March 2008.

Index